STEAM-H: Science, Technology, Engineering, Agriculture, Mathematics & Health

STEAM-H: Science, Technology, Engineering, Agriculture, Mathematics & Health

Series Editor
Bourama Toni
Department of Mathematics
Howard University
Washington, DC, USA

This interdisciplinary series highlights the wealth of recent advances in the pure and applied sciences made by researchers collaborating between fields where mathematics is a core focus. As we continue to make fundamental advances in various scientific disciplines, the most powerful applications will increasingly be revealed by an interdisciplinary approach. This series serves as a catalyst for these researchers to develop novel applications of, and approaches to, the mathematical sciences. As such, we expect this series to become a national and international reference in STEAM-H education and research.

Interdisciplinary by design, the series focuses largely on scientists and mathematicians developing novel methodologies and research techniques that have benefits beyond a single community. This approach seeks to connect researchers from across the globe, united in the common language of the mathematical sciences. Thus, volumes in this series are suitable for both students and researchers in a variety of interdisciplinary fields, such as: mathematics as it applies to engineering; physical chemistry and material sciences; environmental, health, behavioral and life sciences; nanotechnology and robotics; computational and data sciences; signal/image processing and machine learning; finance, economics, operations research, and game theory.

The series originated from the weekly yearlong STEAM-H Lecture series at Virginia State University featuring world-class experts in a dynamic forum. Contributions reflected the most recent advances in scientific knowledge and were delivered in a standardized, self-contained and pedagogically-oriented manner to a multidisciplinary audience of faculty and students with the objective of fostering student interest and participation in the STEAM-H disciplines as well as fostering interdisciplinary collaborative research. The series strongly advocates multidisciplinary collaboration with the goal to generate new interdisciplinary holistic approaches, instruments and models, including new knowledge, and to transcend scientific boundaries.

More information about this series at http://www.springer.com/series/15560

Martha Refugio Ortiz-Posadas

Editor

Pattern Recognition Techniques Applied to Biomedical Problems

 Springer

Editor
Martha Refugio Ortiz-Posadas
Electrical Engineering Department
Universidad Autónoma
Metropolitana-Iztapalapa
Mexico City, Mexico

ISSN 2520-193X ISSN 2520-1948 (electronic)
STEAM-H: Science, Technology, Engineering, Agriculture, Mathematics & Health
ISBN 978-3-030-38023-6 ISBN 978-3-030-38021-2 (eBook)
https://doi.org/10.1007/978-3-030-38021-2

Mathematics Subject Classification: 65D18, 92Bxx, 68T10, 68T30, 68T35

This Springer imprint is published by the registered company Springer Nature Switzerland AG.
The registered company address is: Gewerbestrasse 11, 6330 Cham, Switzerland

Preface

Pattern recognition is the science that studies the processes of identification, characterization, classification, and reconstruction of sets of objects or phenomena, with the purpose of extracting information that allows establishing common properties among them. It has an applied and multidisciplinary character, and it is conformed to technical sciences, computer science, and mathematics, among others, in order to develop computational tools and methodologies related to these processes. The fundamental problems of pattern recognition refer to those related to the determination of factors that affect objects and their classification, and four types are considered: (1) selection of variables and objects, (2) supervised classification, (3) unsupervised classification, and (4) partially supervised classification.

On the other hand, there are different study areas of pattern recognition such as image processing, signal processing, computer vision, remote sensing, neural networks, genetic algorithms, artificial intelligence techniques, descriptive geometry, mathematical morphology, statistical recognition, structural syntactic recognition, and combinatorial logical recognition, to name a few. All these can be applied into biomedical problems.

This volume presents together leading Latin American researchers from five different countries, namely, Brazil, Chile, Costa Rica, México, and Uruguay, to present their own work with the perspective to advance their specific fields. It presents nine chapters regarding different pattern recognition technics applied to the solution of several biomedical problems featured as follows.

In Chap. 1, Aída Jiménez-González and Norma Castañeda-Villa present two experiences on the recovery of physiological information from noisy datasets, applying the method of independent component analysis (ICA).

In Chap. 2, Verónica Jacinto Jiménez et al. describe the identification process of genomic variants and genetic expression profiles for the diagnostic of diseases using high-throughput sequencing methodologies.

In Chap. 3, Leticia Vega-Alvarado et al. propose a system for the automatic detection of the parasite causing Chagas disease in stained blood smears images.

In Chap. 4, Alfonso Rosales-López and Rosimary Terezinha de Almeida propose the use of intervention analysis on time series, using the Box and Tiao approach, as a method for health technology assessment on public health interventions.

In Chap. 5, Millaray Curilem et al. evaluate the possibility of detecting the presence of nausea in chemotherapy patients by processing the electrogastrogram signal.

In Chap. 6, Letícia M. Raposo et al. describe the random forest algorithm, showing an application to predict HIV-1 drug resistance.

In Chap. 7, Luis Jiménez-Ángeles et al. describe an overview of the medical imaging modalities most frequently used for assessment of the cardiac contraction pattern.

In Chap. 8, Franco Simini et al. describe two automatic systems related with home care and personal devices.

In Chap. 9, Tlazohtzin Mora-García et al. propose an evaluation tool based on multi-criteria decision analysis (MCDA) for the replacement of older medical equipment installed at hospitals.

Mexico City, Mexico   Martha Refugio Ortiz-Posadas

Contents

Contributors

Rosimary Terezinha de Almeida Programa de Engenharia Biomédica, COPPE, Universidade Federal do Rio de Janeiro, Rio de Janeiro, Brazil

Pablo Álvarez-Rocha Universidad de la República, Montevideo, Uruguay

Alberto Caballero-Ruiz Instituto de Ciencias Aplicadas y Tecnología, Universidad Nacional Autónoma de México, Mexico City, Mexico
National Laboratory for Additive and Digital Manufacturing (MADiT), Mexico City, Mexico

Norma Castañeda-Villa Electrical Engineering Department, Universidad Autónoma Metropolitana-Iztapalapa, Mexico City, Mexico

Max Chacón Departamento de Ingeniería Informática, Universidad de Santiago de Chile, Santiago, Chile

Millaray Curilem Centro de Física e Ingeniería para la Medicina (CFIM), Universidad de La Frontera, Temuco, Chile

Mariano Flores Hospital Regional Dr. Hernán Henríquez Aravena, Temuco, Chile

Matías Galnares Universidad de la República, Montevideo, Uruguay

Laura Gómez-Romero Unidad de Servicios Bioinformáticos, Instituto Nacional de Medicina Genómica, Mexico City, Mexico

Francisco Heredia-López Centro de Investigaciones Regionales Dr. Hideyo Noguchi, Universidad Autónoma de Yucatán, Mérida, Mexico

Luis Jiménez-Ángeles Departamento de Ingeniería en Sistemas Biomédicos, Universidad Nacional Autónoma de México, Mexico City, Mexico

Aída Jiménez-González Electrical Engineering Department, Universidad Autónoma Metropolitana-Iztapalapa, Mexico City, Mexico

Verónica Jiménez-Jacinto Instituto de Biotecnología, Universidad Nacional Autónoma de México, Mexico City, Mexico

Richard Low Infor-Med Medical Information Systems Inc., Woodland Hills, CA, USA

Verónica Medina-Bañuelos Electrical Engineering Department, Universidad Autónoma Metropolitana-Iztapalapa, Mexico City, Mexico

Carlos-Francisco Méndez-Cruz Centro de Ciencias Genómicas, Universidad Nacional Autónoma de México, Mexico City, Mexico

Tlazohtzin R. Mora-García Electrical Engineering Department, Universidad Autónoma Metropolitana-Iztapalapa, Mexico City, Mexico

Flavio F. Nobre Programa de Engenharia Biomédica, Universidade Federal do Rio de Janeiro, Rio de Janeiro, Brazil

Gabriela Ormaechea Universidad de la República, Montevideo, Uruguay

Martha Refugio Ortiz-Posadas Electrical Engineering Department, Universidad Autónoma Metropolitana-Iztapalapa, Mexico City, Mexico

Fernanda Piña-Quintero Service of Electro-Medicine, National Institute of Pediatrics, Mexico City, Mexico

Letícia M. Raposo Programa de Engenharia Biomédica, Universidade Federal do Rio de Janeiro, Rio de Janeiro, Brazil

Paulo Tadeu C. Rosa Programa de Engenharia Biomédica, Universidade Federal do Rio de Janeiro, Rio de Janeiro, Brazil

Alfonso Rosales-López Gerencia de Infraestructura y Tecnología, Caja Costarricense de Seguro Social, San José, Costa Rica

Leopoldo Ruiz-Huerta Instituto de Ciencias Aplicadas y Tecnología, Universidad Nacional Autónoma de México, Mexico City, Mexico

Hugo Ruiz-Piña Centro de Investigaciones Regionales Dr. Hideyo Noguchi, Universidad Autónoma de Yucatán, Mérida, Mexico

Alejandro Santos-Díaz Departamento de Mecatrónica, Escuela de Ingenieria y Ciencias, Tecnológico de Monterrey, Campus Ciudad de México, Mexico City, Mexico

Gabriela Silvera Universidad de la República, Montevideo, Uruguay

Franco Simini Universidad de la República, Montevideo, Uruguay

Sebastián Ulloa Departamento de Ingeniería Eléctrica, Universidad de La Frontera, Temuco, Chile

Raquel Valdés-Cristerna Electrical Engineering Department, Universidad Autónoma Metropolitana-Iztapalapa, Mexico City, Mexico

Leticia Vega-Alvarado Instituto de Ciencias Aplicadas y Tecnología, Universidad Nacional Autónoma de México, Mexico City, Mexico

Claudio Zanelli Onda Corporation, Sunnyvale, CA, USA

The Classification of Independent Components for Biomedical Signal Denoising: Two Case Studies

Aída Jiménez-González and Norma Castañeda-Villa

Abstract This chapter presents two experiences on the recovery of biomedical signals of interest from noisy datasets, i.e., the extraction of the fetal phono-cardiogram from the single-channel abdominal phonogram and the recovery of the Long Latency Auditory Evoked Potential from the multichannel EEG (in children with a cochlear implant). These by implementing denoising strategies based on (1) the separation of components statistically independent by using Independent Component Analysis (ICA) and, of especial interest in this chapter, (2) the classification of the components of interest by taking advantage of properties such as temporal structure, frequency content, or temporal and spatial location. Results of these two case studies are presented on real datasets, where either focused (1) on rhythmic physiological events such as the fetal heart sounds or (2) on spatially localized events like the cochlear implant artifact, the classification stage has been fundamental on the performance of the denoising process and thus, on the quality of the retrieved signals.

Keywords Blind source separation · Cochlear implant artifact · Fetal heart rate · Independent component analysis · TDSep

1 Introduction

During the last three decades, Independent Component Analysis (ICA) has been recognized as a powerful solution to the matter of revealing the driving forces that underlie a set of observed phenomena [1–3]. Particularly, ICA has been important in the field of biomedical signal processing, where the recovery of very low-amplitude signals from a set of mixtures has posed a challenge that traditional approaches

A. Jiménez-González (✉) · N. Castañeda-Villa
Electrical Engineering Department, Universidad Autónoma Metropolitana-Iztapalapa,
México City, México
e-mail: aidaj@xanum.uam.mx

© Springer Nature Switzerland AG 2020   1
M. R. Ortiz-Posadas (ed.), *Pattern Recognition Techniques Applied to Biomedical
Problems*, STEAM-H: Science, Technology, Engineering, Agriculture,
Mathematics & Health, https://doi.org/10.1007/978-3-030-38021-2_1

like digital filtering do not manage to solve. This has been possible because of the development of multiple and efficient implementations of ICA that make it suitable to separate data that used to be thought as too difficult to process [4–8]. As a result, ICA has been successfully used for the extraction of information of interest from electroencephalographic [5, 9, 10], electrocardiographic [9, 11], and, more recently, electromyographic [12], magnetocardiographic [13, 14], magnetoencephalographic [15, 16], and phonographic recordings [17–20].

Results have been promising and have motivated further research into the development of new ICA algorithms and methodologies for performance evaluation that, while paying attention on high-quality separation (and evaluation) of the independent components (ICs) finally recovered, leave the researcher the responsibility of giving physical (or physiological) sense to them and thus to identify and select the components of interest. The task may sound easy at first, but due to the typical high dimensionality of the datasets (e.g., 7, 16, 19, 30, 36, 49, 50, 55, 64, 68, 128, or 130 channels and different recording lengths) and the usually unknown characteristics of the undesirable sources (both physiological and nonphysiological), it requires objective and, if possible, automatic alternatives that ease what we will refer to as the third step of the procedure for denoising biomedical signals by means of ICA, i.e., the classification of the ICs.

At present, different strategies have been presented in the literature, all dependent on the dataset and, certainly, on the experience of the authors, which illustrates the challenge behind the classification task for biomedical signal denoising. Here, we present our experience on two different sceneries: the extraction of the fetal heart sounds from the single-channel abdominal phonogram (i.e., by classifying rhythmic ICs associated to the fetal cardiac activity) and the elimination of the cochlear implant (CI) artifact from the multichannel EEG (i.e., by classifying independent components with artifactual activity synchronized with the acoustic stimulus and spatially located close to the cochlear implant position). To this end, this chapter is organized as follows: Section 2 presents the theoretical background of ICA, while Sects. 3 and 4, respectively, detail the sceneries for denoising the abdominal phonogram and the EEG along with their results and discussion. Finally, Sect. 5 presents our final remarks.

2 Independent Component Analysis

ICA is a statistical algorithm whose aim is to represent a set of mixed signals as a linear combination of statistically independent sources. This technique estimates ICs, denoted by $\hat{s}(t)$, from a group of observations, $x(t)$, which are considered as linear and instantaneous mixtures of unknown sources, $s(t)$ [21]. The statement that in a biomedical signal different sensors/electrodes receive different mixtures of the sources is exploited by ICA, that is, spatial diversity. Spatial diversity means that ICA looks for structures across the sensors and not (necessarily) across time.

ICA identifies the probability distribution of the measurements, given a sample distribution [22].

In the most simplistic formulation of the ICA problem (noise-free), p measured signals $\mathbf{x}(t)$ are a linear mixture of unknown but statistically independent q sources $\mathbf{s}(t)$; each source has moments of any order, zero mean, and $p \geq q$. Then, the ICA model is as follows:

$$\mathbf{x}(t) = \mathbf{A}\,\mathbf{s}(t), \tag{1}$$

where the square mixing matrix $\mathbf{A}$ is also unknown but invertible, $\mathbf{x}(t) = [x_1, x_2, \ldots, x_p]^T$ and $\mathbf{s}(t) = [s_1, s_2, \ldots, s_q]^T$. ICA calculates the demixing matrix, $\mathbf{W} = \mathbf{A}^{-1}$, from the observations $\mathbf{x}(t)$ and estimates the original sources by a linear transform:

$$\hat{\mathbf{s}}(t) = \mathbf{W}\,\mathbf{x}(t), \tag{2}$$

where $\mathbf{W}$ is found by maximizing the statistical independence of the output components [23].

Currently, depending on the method used to seek statistical independence (i.e., higher-order statistics or time-structure-based algorithms), different ICA implementations have been developed. Among the wide range of ICA algorithms, three are frequently used in the field of biomedical signal processing: FastICA [24], Infomax [25, 26], and Temporal Decorrelation Source Separation (TDSep) [27]. The theory behind these three ICA algorithms will be explained in the next sections, all of them share steps (a) and (b) as the initial stage:

(a) *Centering*: Subtract the mean of the mixtures, which simplifies the ICA algorithm $\mathbf{x}(t) = \mathbf{x}(t) - E\{\mathbf{x}(t)\}$, where $E\{\mathbf{x}(t)\}$ is the mean vector of the measurements; when the algorithm is finished, the mean vector is added back.
(b) *Whitening or sphering*: In this preprocessing step, the covariance matrix is calculated as $\mathbf{R}_{\mathbf{xx}} = E\{\mathbf{x}(t)\mathbf{x}^T(t)\}$, and then, an eigenvalue decomposition is performed on it; the decomposition is given by $\mathbf{R} = \mathbf{E}\Lambda\mathbf{E}^T$, where $\mathbf{E}$ is the orthonormal matrix of eigenvectors of $\mathbf{R}$, and Λ is the diagonal matrix of eigenvalues. Transforming the covariance matrix into an identity matrix, a whitening $\mathbf{M}$ matrix is calculated as $\mathbf{M} = (\Lambda^{1/2}\mathbf{E}^T)^{-1}$. This is also known as a principal component decomposition.

2.1   FastICA

This is a computationally efficient algorithm that uses simple estimators of negative entropy, $J(y)$, to search a $\mathbf{W}$ matrix that, when applied to mixtures, maximizes this property in the resulting components, thus allowing the estimation of sources with non-Gaussian probability distributions [28, 29]. It estimates ICs by following either the deflation approach (Defl), where the components are extracted one by one, or the symmetric approach (Sym), where the components are simultaneously extracted. The negative entropy is defined as follows:

$$J(w) = \left[E\left\{ G\left(w^T v \right) \right\} - E\left\{ G(v) \right\} \right]^2,$$

(3)

where w is an m-dimensional vector such as $E\{(w^T v)^2\} = 1$, v is a Gaussian variable with zero mean and unit variance, and G is a nonquadratic cost function, e.g., tan$h$ or y^3. The problem is now reduced to find a transformation $\mathbf{W}$ whose vectors, w, are iteratively adjusted to maximize J which is equivalent to reduce the mutual information (MI); this is performed by a fixed-point algorithm. From choosing an initial weight vector w, the algorithm calculates the direction of w maximizing the non-Gaussianity of the projection $w^T x$ (linear combination of the measured signals). Since the signal is already whitened, to make the variance of $w^T x$ unity, it is sufficient to constrain the norm of the pseudo-inverse of the initial weight vector w^+, to be unity, $w = w^+/\|w^+\|$; if the old and new values of w do not point in the same direction, the algorithm recalculates the direction of w. Finally, the demixing matrix is given by $\mathbf{W} = w^T \mathbf{M}$ and the estimations $\hat{\mathbf{s}}(t)$ by Eq. 2.

2.2   Infomax and Ext-Infomax

Described by Bell and Sejnowski [25], Infomax is an ICA algorithm which uses the mutual information between the estimated sources as a criterion of the minimization of independence, with which the joint negative entropy is maximized. The demixing matrix $\mathbf{W}$, which is found using stochastic gradient ascent, maximizes the entropy of an input vector $\mathbf{x_G}$, linearly transformed $\mathbf{u} = \mathbf{W}\mathbf{x_G}$, and sigmoidally compressed $y = g(\mathbf{u})$. Then, $\mathbf{W}$ performs component separation while the nonlinear $g(.)$ provides the necessary high-order statistic information, $g(\mathbf{u}_i) = (1 + \exp(-\mathbf{u}_i))^{-1}$. This gives an update rule $\hat{\mathbf{u}}_i = 1 - 2\mathbf{u}_i$. *Infomax* is able to decompose signals into ICs with sub- and super-Gaussian distributions in its extended version. The original learning rule for super-Gaussian distributions is as follows:

$$\Delta \mathbf{W} \propto \left[\mathbf{I} - \tan h(\mathbf{u})\, \mathbf{u}^T - \mathbf{u}\mathbf{u}^T \right] \mathbf{W},$$

(4)

where $\mathbf{I}$ is the identity matrix, and $\mathbf{u}$ are the estimated sources. The extended learning rule (Ext-Infomax) [26] for sub-Gaussian distributions is as follows:

$$\Delta \mathbf{W} \propto \left[\mathbf{I} - \Lambda \tan h(\mathbf{u})\, \mathbf{u}^T - \mathbf{u}\mathbf{u}^T \right] \mathbf{W}.$$

(5)

The algorithm switches between two learning rules: one for sub-Gaussian and one for super-Gaussian sources. Λ is a diagonal matrix which includes the switching criterion between the two learning rules: $\Lambda_{ij} = 1$ for super-Gaussian and -1 for sub-Gaussian. Finally, the estimated sources $\hat{s}(t)$ are computed by Eq. 2.

2.3 TDSep

The Temporal Decorrelation Source Separation (TDSep) algorithm [27] takes into account the temporal structure of the signals. TDSep uses several time-delayed second-order correlation matrices for source separation. JADE and TDSep determinate the mixing matrix based on a joint approximate diagonalization of symmetric matrices; the principal difference between these two algorithms is that JADE maximizes the kurtosis of the signals, while TDSep minimizes temporal cross-correlation between the signals [30].

TDSep could be summarized as follows: firstly, Ziehe and Müller define a cost function based on a certain time lag τ and a time average that measure the correlation between the signals; after whitening, this cost function imposes decorrelation over time. After that, they propose an alternative technique for the joint diagonalization using a rotation [31]. In the rotation step, the cost function can be minimized by approximate simultaneous diagonalization of several correlation matrices through several elementary JACOBI rotations [22] to obtain the so-called rotation matrix, $\mathbf{Q}$.

TDSep computes those matrices relying only on second-order statistics and diagonalizes the covariance matrices $\mathbf{R_{xx}}$ for a time lag $\tau = 0$ and at the same time diagonalizes the covariance matrix for a given delay $\mathbf{R_{x\tau}} = E\{\mathbf{x}(t)\mathbf{x}(t - \tau)^T\}$. The source covariance matrix $\mathbf{R_{s\tau}}$ is diagonal for all time lags $\tau = 0, 1, 2,$ $\dots, N-1$ as $\mathbf{R_{s\tau}} = \mathbf{WR_{x\tau}W}^T$, where $\mathbf{R_{x\tau}}$ is the signal covariance matrix. This algorithm determines the mixing matrix based on a joint approximate digitalization of symmetric matrices. Finally, using the whitening matrix $\mathbf{M}$ and the rotation matrix $\mathbf{Q}$, an estimate of the mixing matrix can be calculated as follows:

$$\hat{\mathbf{A}} = \mathbf{M}^{-1}\mathbf{Q}, \tag{6}$$

and the estimations $\hat{s}(t)$ by Eq. 2.

3 Case Study I: Denoising the Abdominal Phonogram

This first case describes a methodology for the recovery of the fetal heart sounds (i.e., the fetal phonocardiogram, FPCG) from the single-channel abdominal phonogram, which has been implemented by using a denoising strategy based on what has been referred to as single-channel ICA (SCICA) [17, 20, 32].

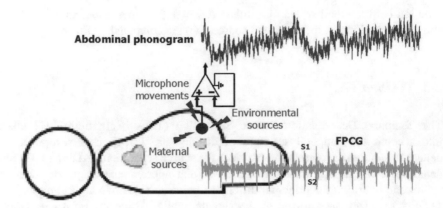

Fig. 1 Distortion of the FPCG by physiological and nonphysiological sources during the abdominal recording. The FPCG corresponds to the acoustic vibrations produced by the fetal heart, where the two main heart sounds (S1 and S2) are indicated. The abdominal phonogram, which is recorded by a microphone positioned on the maternal womb, is a mixture of multiple sources, where the fetal information is hardly noticed

3.1   The Problem Definition

During the last two decades, fetal examination by passive detection of cardiac vibrations has regained attention [17, 33–38]. The technique is performed by positioning a sensitive acoustic sensor on the maternal womb to record the abdominal phonogram, a signal that is rich in information for well-being assessment (i.e., it contains the two main fetal heart sounds S1 and S2, which can be used to estimate the heart rate and the heart valve condition) but highly attenuated by the amniotic fluid and abdominal tissues and, consequently, highly immersed (both temporally and spectrally) in maternal and environmental sources whose characteristics turn the recovery of fetal information into a challenging task. Fig. 1 illustrates the acoustic vibrations produced by the beating fetal heart (i.e., the FPCG, as a free of noise trace where the two main heart sounds, S1 and S2, can be detected) and the acoustic vibrations actually recorded at the surface of the maternal womb (i.e., the abdominal phonogram, where the fetal information is hard to observe).

In our work, as an alternative to the rigid approach given by traditional digital filtering schemes (which assume that the FPCG spectrum does not change, independently on the fetal age, heart rate, or condition), we have studied a data-driven strategy that, by taking advantage of the rich time-structure in the abdominal phonogram, adapts itself to each dataset to extract the fetal heart sounds from the abdominal phonogram (i.e., the FPCG) [17]. This approach, known as SCICA, has been applied to a set of 25 noisy single-channel abdominal phonograms. This first case study will describe such a dataset in Sect. 3.2, detail the implementation of SCICA for denoising purposes in Sect. 3.3, present our findings in Sect. 3.4, and, finally, a brief Discussion in Sect. 3.5.

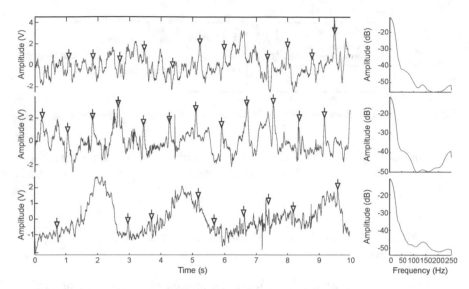

Fig. 2 Time (left-hand side) and frequency (right-hand side) representations of three abdominal phonograms in the dataset. (Modified from [39])

3.2 A Dataset of Single-Channel Abdominal Phonograms

The dataset is composed of 25 single-channel recordings that were digitized at a sampling frequency of 500 Hz during 3 or 5 minutes (MP100, Biopac Systems™). The data was obtained from 18 pregnant women (24 ± 3 years old and fetal gestational ages between 29 and 40 weeks, who provided their informed consent to participate in the study) by using a PCG piezoelectric transducer (TK-701T, Nihon Kohden™) connected to a general purpose amplifier (DA100, Biopac Systems™). Additionally, as a reference signal, the abdominal ECG was simultaneously recorded. Figure 2 presents segments (10 s length) of three abdominal phonograms in the dataset and their frequency content, where the signals clearly show a slow component along with some quasiperiodic peaks (indicated by downward arrows), but without any clear evidence of the FPCG.

3.3 Single-Channel ICA (SCICA)

The implementation of SCICA requires three stages [17, 32]: a preprocessing stage that projects the single-channel abdominal phonogram into a higher dimensional space, a processing stage that transforms such a multichannel representation into a set of multiple ICs, and, finally, a postprocessing stage that, in this case study, automatically classifies the rhythmic ICs corresponding to the fetal cardiac activity

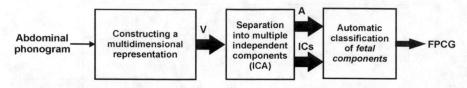

Fig. 3 Denoising procedure to extract the FPCG from the abdominal phonogram by using SCICA

and, thus, makes it possible to estimate the source of interest, i.e., the FPCG. These three stages are illustrated in Fig. 3 and described in Sects. 3.3.1, 3.3.2 and 3.3.3, respectively.

3.3.1 Mapping a Single-Channel Signal into a Multidimensional Representation

This first stage is implemented by using the method of delays (MD or dynamical embedding), which makes it possible to project an N_T-scalar time series $\{x_i\}_{i=1,\ldots,N_T}$ into a multichannel representation of the data [40]. This new representation, also known as the matrix of delays ($\mathbf{V}$), is built up by taking consecutive delay vectors of length m as $v_k = \{x_k, x_{k+1}, \ldots, x_{k+m-1}\}_{k=1,\ldots,N}$ and by using them as column vectors in the m-dimensional matrix $\mathbf{V} = \begin{bmatrix} v_1^T, v_2^T, \ldots, v_N^T \end{bmatrix}$. There, N corresponds to the number of delay vectors ($N = N_T - (m-1)$), and m corresponds to the embedding dimension ($m \geq fs/fl$), where fs is the sampling frequency (i.e., $fs = 500$ Hz) and fl is the lowest frequency component of interest in x_i.

In our work, by considering that the component with the lowest frequency in the FPCG is 10 Hz [41, 42], the value used for m has been 50. Regarding N, we empirically found that $N_T = 5000$ samples (i.e., 10 s) was good enough to ensure that the matrix of delays covered a quasi-stationary signal and then that ICA would converge. Thus, by using these parameters, we are certain that the multidimensional representation $\mathbf{V}$ ($m \times N$) is rich in temporal information about the fetal cardiac activity and, most importantly, that it is ready to be spanned by a convenient basis such as ICA.

3.3.2 Extraction of Multiple Independent Components

As mentioned in Sect. 2, a number of ICA implementations is currently available, and since the key point in each algorithm is the method used to numerically calculate statistical independence (e.g., higher-order statistics or time-structure-based algorithms), their performance on the ICs estimation will be different. In our research [19], we have studied two ICA implementations and found that the underlying components in the abdominal phonogram are better recovered by using the Temporal Decorrelation Separation approach of TDSep [27] rather than by using the approach of FastICA [43].

In general words, the implementation of TDSep (1) defines independence by the absence of cross-correlations among the underlying components and (2) assumes that such components possess some temporal structure that, consequently, produce diagonal time-delayed correlation matrices, R_τ^V. Hence, TDSep analyzes the dependency structure of the multichannel representation, V, by creating a set of square matrices and then by finding the joint diagonalizer of that set, which turns out to be the mixing matrix ($\hat{A}$) mentioned in Eq. 6. To this end, TDSEP calculates a set of time-lagged correlation matrices of V by $R_\tau^V = E\{V[n]V[n+\tau]\}$, where E represents expectation and τ ($= 0, 1, 2, \ldots, k$) is the time lag. Then, since for independent components these matrices have to be diagonal, TDSep performs a joint diagonalization of R_τ^V to estimate A. Finally, after calculating W (which is the inverse of A), it is possible to substitute it in Eq. 2 to estimate the constituent ICs in our matrix of delays V.

In our research, knowing that the value of k defines the number of time lags and the quality of the separate ICs, we tested different k values and found that, in our dataset of abdominal phonograms, a value of $k = 1$ makes is possible to consistently process V and estimate the constituent ICs. As a result, TDSep estimates m ICs whose typical spectra are given by a well-defined single peak [19]. These ICs must be further analyzed to complete the denoising process as detailed in Sect. 3.3.3.

3.3.3 Automatic Classification of Fetal Independent Components

According to [32], when applying ICA to the matrix of delays V, some of the estimate ICs will belong to the same independent process, which in our case can be described as either physiological or nonphysiological (e.g., the fetal cardiac activity, the maternal cardiac activity or line noise). Here, since we are only interested in retrieving the fetal cardiac information in the form of the FPCG, the denoising process will require identifying and grouping the ICs corresponding to the fetal subspace while discarding the others.

Before describing the classification process in this final stage of SCICA, it is imperative to mention two fundamental and consistent characteristics of the ICs recovered from the abdominal phonograms by TDSep; they are rhythmic and spectrally band-limited components as mentioned in [19, 44] and studied in [45]. These two characteristics are essential for this section, which describes a methodology for the automatic classification of the fetal cardiac ICs by using a couple of spectral features that are relatively easy to calculate, i.e., their rhythmicity and spectral content [20, 44, 45].

The procedure for this stage is conducted in four steps as follows:

(a) *Projecting ICs back to the measurement space* [17]: In this step, each IC is processed by $Y^i = a_i c_i^T$, with a_i as the ith-column in the mixing matrix A (obtained by TDSep), c_i as the ith-IC, and Y^i as the resulting matrix of delays for that IC ($i = 1, 2, \ldots, 50$). Next, $\mathbf{Y}^i$ is hankelized to produce the ith-projected IC (IC_p^i), by $\mathrm{CI}_p^i = \frac{1}{50} \sum_{k=1}^{50} \mathbf{Y}_{k,(t+k-1)}^i$, where $t = 1, 2, \ldots, N$ [46].

(b) *Calculating spectral features*: This step quantifies the rhythmicity and frequency content of each IC_p^i as illustrated in Fig. 4 and described in the next paragraphs.

   (i) *Frequency content of an IC_p^i (S_I)*: This index is used as an indicator of the spectral content of the IC_p^i under analysis and is calculated using the Welch's method, with a Hanning window of 32 samples and an overlap of 50%. Then, from the characteristic band-limited spectrum obtained, $\hat{S}_x$, the frequency of its peak is taken as the index that represents the frequency content, S_I. Figure 4 presents an example of the S_I calculation for IC_p^{45} (one of the independent components presented in Sect. 3.4). The resulting trace, as can be observed, is band-limited and presents a well-defined single peak that is centered at $S_I = 27$ Hz.

   (ii) *Rhythmicity of an IC_p^i (R)*: This index is used as an indicator of the physiological generator driving the IC_p^i. As illustrated in Fig. 4, the calculation of R starts by generating a normalized envelope of the IC_p^i using the Hilbert transform. Next, the envelope autocorrelogram is generated and band-pass filtered from 0.7 to 3.1 Hz by a 10-order FIR filter (to reduce contributions from the maternal respiratory rhythm and from the harmonics of the maternal and fetal cardiac rhythms, respectively). Then, the filtered autocorrelogram is transformed into the frequency domain using the Welch's periodogram, with a Hanning window of 2018 samples and 50% overlap (which contains an average of eight fetal heart beats). Finally, from the resulting autospectrum, $\hat{S}_{xx}$, the frequency of the largest peak is used as the rhythmicity indicator, R. Figure 4 presents an example of the R calculation for IC_p^{45}. The resulting trace, as can be observed, presents a single rhythm that is centered at $R = 2.3$ Hz.

(c) *Selecting fetal components*: This third step starts by verifying whether the indexes R and S_I are between 1.7 and 3.0 Hz for the former and between 19.0 and 44.5 Hz for the latter, which are the intervals that our research has empirically identified as typical for the IC_p^is corresponding to the fetal cardiac process [44, 45]. In those cases where R and S_I disagree on the IC_p^i category, the S_I value has the highest priority, which means that, if S_I is in the fetal interval, then the component is classified as fetal cardiac without any further analysis. Alternatively, whenever both indexes R and S_I point at the component as a fetal IC_p^i, it is further analyzed to verify its stability (P) and identify if it is driven by a single biological rhythm (i.e., the fetal heart rate) or if it is contaminated by a second one (typically the maternal heart rate). Then, if the IC_p^i is found to be a stable component, it is classified as a fetal cardiac component. On the contrary, in those cases where the IC_p^i is found to be contaminated by a second biological rhythm (which happens to be the cardiac maternal rhythm), it is further analyzed to establish the level of contamination, and only the

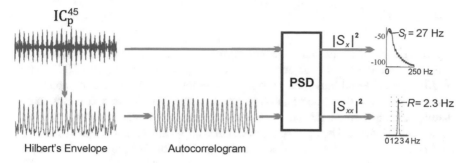

Fig. 4 Calculation of two spectral indexes (rhythmicity (R) and frequency content (S_I)) for automatic classification of the fetal cardiac components extracted by SCICA

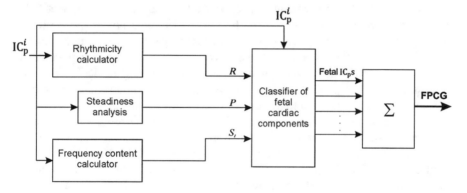

Fig. 5 Procedure for the identification and automatic classification of fetal cardiac components and construction of the FPCG trace in the third stage of SCICA

components where the contribution of the second rhythm is not important are classified as fetal cardiac components. This is done by calculating two Pearson's coefficients, one between the autocorrelogram and a sinusoid oscillating at the fetal heart rate and another between the autocorrelogram and a sinusoid oscillating at the maternal heart rate. Next, if the former Pearson's coefficient is larger than the latter (and S_I indicates that the component belongs to the fetal cardiac group), it is classified as fetal cardiac. Details of the complete version of this algorithm can be found in [20, 44], along with an evaluation of its performance. In this chapter, we are only describing the procedure that makes it suitable to classify the components that are more likely to belong to the fetal cardiac group.

(d) *Constructing the FPCG*: In this final step, the classified fetal cardiac IC_ps are summed for the construction of the FPCG trace as $FPCG = \sum$ fetal IC_ps [17].

Figure 5 presents the four steps followed in our methodology for classifying fetal IC_ps and constructing the trace corresponding to the FPCG.

3.4   Results

Figure 6 depicts the time and frequency representations of a 10 s segment of a noisy abdominal phonogram of subject 1 (40 weeks of gestational age) and 10 IC_ps (out of 50) separated and classified by our methodology for denoising the abdominal phonogram. From the abdominal phonogram, in the time domain, it is possible to distinguish a low-frequency component (with large amplitude between ± 2 V) and some peaks (indicated by upward arrows), but without any clear evidence of the fetal cardiac activity (i.e., the Fetal Heart Sounds (FHS)). In the frequency domain, the spectrum indicates that most of its power is concentrated below 75 Hz (> -30 dB), and the autospectrum indicates that four rhythms are present, with the strongest rhythm centered at 1.3 Hz.

Regarding the IC_ps, their frequency representations, as mentioned in Sect. 3.3.3, can be described as consistently given by (1) band-limited power spectra (i.e., a single-peak spectra) and (2) well-defined autospectra centered at a single dominant biological rhythm (below 4 Hz). These two characteristics made it possible for SCICA to automatically classify five of them as belonging to the fetal cardiac process (IC^{42}_p, IC^{43}_p, IC^{44}_p, IC^{45}_p, and IC^{46}_p). Such components, in the time domain, showed periodic activity almost every 0.45 s and appeared very clean (except for some disturbances indicated by a downward arrow), whereas in the frequency domain presented a power spectrum centered between 39 Hz (IC^{42}_p) and 21 Hz (IC^{46}_p), all of them with a rhythm centered at 2.3 Hz. The remaining components, IC^{41}_p, IC^{47}_p, IC^{48}_p, IC^{49}_p, and IC^{50}, although also depicted temporal structure (clearly indicated by the upward arrows in IC^{49}_p) and band-limited spectra, do not belong to the fetal cardiac process of interest in this study and, thus, were discarded during the classification stage as part of the denoising strategy followed by SCICA. At the end, as can be seen, among the 50 IC_ps extracted from this segment of abdominal phonogram in the second stage of SCICA, only five components were used to construct the trace corresponding to the fetal cardiac activity, i.e., the FPCG. This example illustrates one of the best cases, where the separate IC_ps were driven by a single biological rhythm (observed in both the temporal structure and autospectrum) that, detected by a "relatively" easy set of rules, successfully classified the fetal components of interest.

Figure 7 illustrates the time and frequency representations of a 10 s segment of a noisy abdominal phonogram of subject 1 (40 weeks of gestational age) and its denoised version (i.e., the FPCG with its two main heart sounds, S1 and S2) generated by SCICA. Complementary, as a reference signal, the abdominal ECG is included. The abdominal phonogram, as can be seen, is a noisy trace composed of different sources that unease the identification of the fetal information in both time and frequency domains. On the contrary, the denoised trace produced by SCICA (i.e., the FPCG) shows periodic activity at about every 0.45 s with amplitude of ± 1 V. In addition, such activity is temporally aligned with the fetal QRS complex in the abdominal ECG (indicated by a dotted vertical line), which confirms that this trace actually corresponds to the FPCG (where the two main heart sounds, S1 and S2, can be seen). Also, the time series shows that S1 has the highest amplitude and

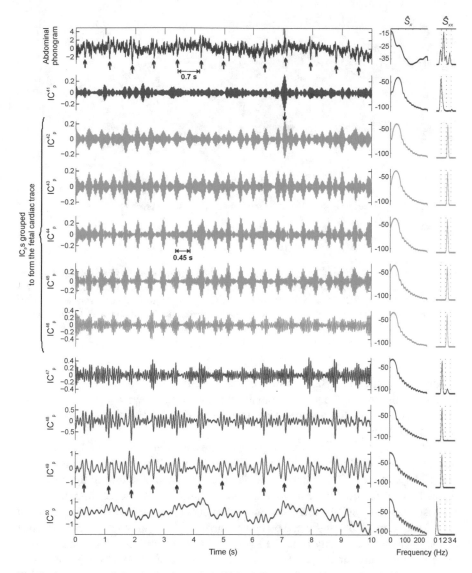

Fig. 6 A segment of the single-channel abdominal phonogram recorded from subject 1 (gestational age = 40 weeks) and 10 IC_ps retrieved by SCICA (out of 50). From left to right: the time series, the power spectrum ($\hat{S}_x$), and the autospectrum ($\hat{S}_{xx}$). IC^{42}_p to IC^{46}_p correspond to five components that were automatically classified as fetal cardiac. The upward arrows in the abdominal phonogram point at some peaks contaminating the signal (also present in IC^{48}_p and IC^{49}_p), whereas the downward arrow points at some residual interference in the fetal components estimated. Units of the y-axis in the time series are V and dBs in the power spectrum. The autospectrum is a normalized signal with unitary amplitude

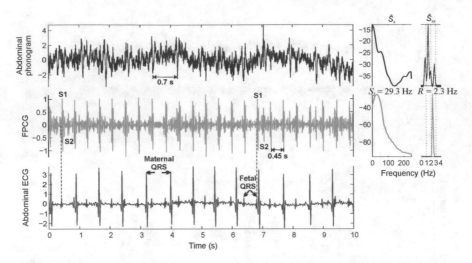

Fig. 7 A segment of the single-channel abdominal phonogram recorded from subject 1 (gestational age = 40 weeks), its denoised version (i.e., the FPCG with its two main heart sounds, S1 and S2) generated by SCICA, and the abdominal ECG (as a reference signal). From left to right: the time series, the power spectrum ($\hat{S}_x$), and the autospectrum ($\hat{S}_{xx}$) along with the indexes of frequency content (S_I) and rhythmicity (R). Units of the y-axis in the time series are V and dBs in $\hat{S}_x$. $\hat{S}_{xx}$ is a normalized signal with unitary amplitude

is always present, while S2 has lower amplitude and is hard to detect in some heart cycles. Finally, in the frequency domain, the spectrum shows a well-defined peak centered at 29.3 Hz and a single biological rhythm at 2.3 Hz (that agrees with the fetal heart rate).

Figure 8 depicts the time and frequency representations of another 10 s segment of a noisy abdominal phonogram, this time of subject 3 (38 weeks of gestational age) and 10 IC_ps (out of 50) separated and classified by our methodology for denoising the abdominal phonogram. From the abdominal phonogram, in the time domain, it is possible to distinguish a low-frequency component (with large amplitude between ± 2 V) and some peaks (indicated by upward arrows), but without any clear evidence of the fetal cardiac activity (i.e., the FHS). In the frequency domain, the spectrum indicates that most of its power is concentrated below 75 Hz (> -30 dB), and the autospectrum indicates that two rhythms are present, with the strongest rhythm centered at 0.9 Hz.

Regarding the IC_ps, while their frequency spectra can be observed as consistently given by (1) band-limited traces (i.e., a single-peak spectra), their (2) autospectra, contrary to those in subject 1, were not always centered at a single biological rhythm (as clearly seen in IC^{46}_p, IC^{48}_p, and IC^{50}_p, where two rhythms are present). As a result of the characteristics of $\hat{S}_x$ and $\hat{S}_{xx}$ in this example, SCICA automatically classified four components as belonging to the fetal cardiac process (IC^{42}_p, IC^{43}_p, IC^{44}_p, and IC^{45}_p). In the time domain, these components showed periodic activity at almost every 0.37 s and appeared clean (except for some disturbances indicated by

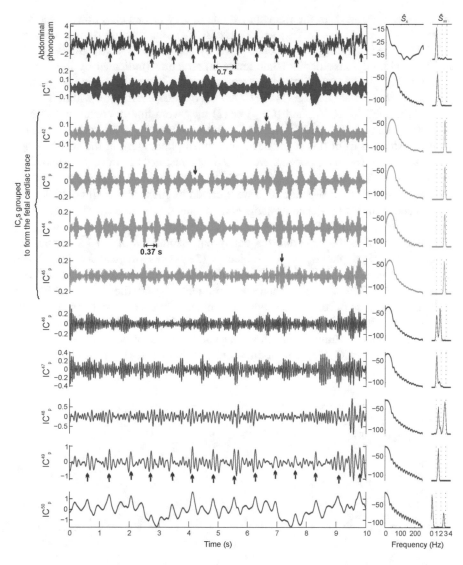

Fig. 8 A segment of the single-channel abdominal phonogram recorded from subject 3 (gestational age = 38 weeks) and 10 IC_ps retrieved by SCICA (out of 50). From left to right: the time series, the power spectrum ($\hat{S}_x$), and the autospectrum ($\hat{S}_{xx}$). IC^{42}_p to IC^{45}_p correspond to four components that were automatically classified as fetal cardiac. The upward arrows in the abdominal phonogram point at some peaks contaminating the signal (also present in IC^{49}_p), whereas the downward arrow points at some residual interference in the fetal components estimated. Units of the y-axis in the time series are V and dBs in the power spectrum. The autospectrum is a normalized signal with unitary amplitude

the downward arrows). In the frequency domain, they presented a power spectrum centered between 41 Hz (IC^{42}_p) and 27 Hz (IC^{45}_p) and a single rhythm centered at 2.7 Hz. The remaining components, given by IC^{41}_p, IC^{46}_p, IC^{47}_p, IC^{48}_p, IC^{49}_p, and IC^{50}, although also depict some temporal structure (indicated by the upward arrows in IC^{49}_p) and band-limited spectra, were not taken as belonging to the fetal cardiac process either because (1) they were not driven by a rhythm within the fetal interval (e.g., IC^{41}_p, IC^{47}_p, and IC^{49}_p) or (2) they were highly contaminated by a second rhythm (e.g., IC^{46}_p, IC^{48}_p, and IC^{50}_p). Consequently, in this case, among the 50 IC_ps extracted from this segment of abdominal phonogram in the second stage of SCICA, only four components were used to construct the trace corresponding to the fetal cardiac activity, i.e., the FPCG. Thus, this second example presented a case where some of the separate IC_ps were driven by more than a single biological rhythm and then that produced unstable components that were discarded during the classification stage of SCICA.

Figure 9 depicts the time and frequency representations of a 10 s segment of a noisy abdominal phonogram of subject 3 (38 weeks of gestational age) and its denoised version (i.e., the FPCG with its two main heart sounds, S1 and S2) generated by SCICA. Complementary, as a reference signal, the abdominal ECG is included. As can be observed, the abdominal phonogram is a noisy trace composed of different sources that unease the identification of the fetal information in both time and frequency domains. On the other hand, the denoised trace produced by SCICA (i.e., the FPCG) shows periodic activity at about every 0.37 s with amplitude

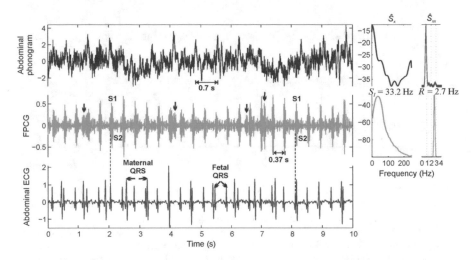

Fig. 9 A segment of the single-channel abdominal phonogram recorded from subject 3 (gestational age = 38 weeks), its denoised version (i.e., the FPCG and its two main heart sounds, S1 and S2) generated by SCICA, and the abdominal ECG (as a reference signal). From left to right: the time series, the power spectrum ($\hat{S}_x$), and the autospectrum ($\hat{S}_{xx}$) along with the indexes of frequency content (S_I) and rhythmicity (R). Units of the y-axis in the time series are V and dBs in $\hat{S}_x$. $\hat{S}_{xx}$ is a normalized signal with unitary amplitude

of ± 0.5 V. Such activity, as mentioned in the former example, is temporally aligned with the fetal QRS complex in the abdominal ECG (indicated by a dotted vertical line), which confirms that this trace actually corresponds to the FPCG (where the two main heart sounds, S1 and S2, can be seen). Again, the time series indicates that S1 has the highest amplitude and is always present, while S2 has lower amplitude and is hard to detect in some heart cycles. Moreover, as indicated by the downward arrows, the FPCG retrieved in this case has some disturbances. Finally, in the frequency domain, the spectrum shows a well-defined peak centered at 29.3 Hz and a single biological rhythm of 2.3 Hz (that agrees with the fetal heart rate).

Similar results were observed after denoising the 25 abdominal phonograms in the dataset, where the FPCGs recovered consistently showed S1 as the heart sound with the largest amplitude and easier to visualize, while S2 was hard to spot in some cases. In the frequency domain, the FPCG spectrum was always characterized by a single and well-defined peak that was centered at 30.5 ± 2.1 Hz and by an autospectrum concentrated at 2.4 ± 0.2 Hz.

3.5 Discussion and Conclusions

This case study has described SCICA as a three-stage methodology for denoising the abdominal phonogram and retrieving a physiological source of interest, the fetal phonocardiogram (FPCG). Such a methodology, developed by combining the method of delays with an implementation of ICA, gives rise to a set of 50 independent components that are further analyzed to automatically classify those belonging to the fetal cardiac process and, then, used for constructing the estimate of the FPCG. The classification task takes advantage of the rich time-structure in the fetal cardiac activity underlying the abdominal phonogram, which is identified by two indexes that are easy and fast to calculate, the frequency content (S_I) and the rhythmicity (R) of the components separated by ICA. Then, after testing the degree of contamination by a secondary rhythm, the classifier decides whether the component under analysis is useful (as a fetal cardiac component) or not. Finally, the classified components are used to construct the trace corresponding to the FPCG.

One of the most important aspects for our classifier to work, as expressed in the two examples presented in Sect. 3.4, is given by the characteristics of the IC_ps. Ideally, the IC_ps must be (1) spectrally disjoint and (2) driven by a single biological rhythm to guarantee a "clean" (free of disturbances) and reliable reconstruction of the FPCG. In our work, we have found that TDSep produces IC_ps that consistently accomplish criterion 1 (as seen in the two examples in Sect. 3.4) but that may fail at achieving criterion 2 (as in the second example in Sect. 3.4). This means that, to improve the outcome of the denoising procedure of the abdominal phonogram conducted by SCICA, further research is needed on the separation stage to ensure the stability of the IC_ps (i.e., testing other ICA implementations to improve the quality of the separation between the maternal and fetal cardiac information) and thus to increase the influence of the classification stage on the quality of the FPCG finally reconstructed (e.g., to improve the signal-to-noise ratio, SNR). Certainly, the three stages are equally important on the performance of SCICA, although this chapter paid attention to the classification stage due to the focus of this book.

So far, the methodology described in this case study has been applied to the recovery of FPCG estimates from 25 noisy single-channel abdominal phonograms. The FPCGs recovered by SCICA from the dataset consistently showed that S1 has the highest amplitude and is always present, while S2 has lower amplitude and is hard to detect in some heart cycles. In the frequency domain, on the other hand, the spectrum shows a well-defined peak centered at 30.5 ± 2.1 Hz and a single biological rhythm centered at 2.4 ± 0.2 Hz (which is in the interval reported as normal for the fetal heart rate [47]). The reasons behind the low amplitude of S2 are unknown, although they could be related to factors like the distance between the sensor and the fetal heart or the path followed by the acoustic signal toward the sensor, which might be attenuating the frequencies corresponding to S2. Regarding the spectral behavior, when comparing to other works, it was found that the authors reported spectral peak values centered at 30 Hz for S1 and at 75–100 Hz for S2 [42], at 23 Hz for S1 and at 19 Hz for S2 [48], and between 37 and 54 Hz for S1 [49], indicating that the values found in our dataset are not far from those reported in the literature.

Finally, proven that a good-quality FPCG is available after denoising the abdominal phonogram (our empirical observations indicate that the SNR should be larger than 20 dB in the signal envelope [50]), there should be suitable to further process the FPCG (1) to measure the beat-to-beat FHR by using S1 (or S2 if its amplitude is large enough) or (2) to perform some morphological analysis of the FHS to study the heart valve condition (a detailed description of such analysis can be found in [50]).

4   Case Study II: Denoising the EEG for Recovering the Late Latency Auditory Evoked Potentials (LLAEP)

4.1   The Problem Definition

An Auditory Evoked Potential (AEP) is the response of different parts of the auditory system elicited by an acoustic stimulus; represent the different relay centers in the auditory via (from the cochlea to the auditory cortex, passing through the auditory nerve) [51]. On analyzing the characteristic of the response, generally amplitude and latencies of the waveforms, it is possible to establish the region or regions in the auditory system which generated the response. Each peak of the waveform represents the synchronous activity of sequential nuclei in the ascending auditory pathway. Since this potential is obtained from coherent average of EEG trials, the common EEG artifacts could be present even in the AEPs, for example, blinking, eye movements, and line noise [52].

EEG recordings at the scalp electrodes are assumed as a linear mixture of the underlying brain sources – AEP, alpha and beta activities, etc. – and the artifact signals – blinking, muscle noise, etc. Additionally, volume conduction in the brain is linear and instantaneous; then, the ICA algorithm is plausible for source separation from EEG data [53, 54].

Specifically, the Long Latency Auditory Evoked Potential (LLAEP) has been used for following changes in latency and morphology of their waves in order to objectively evaluate the maturation of the auditory system, for example, in children with cochlear implants (CIs). Nevertheless, CI can induce an artifact in the EEG/LLAEP recordings when sounds are presented, which makes the analysis of this potentials much harder; ICA has been used to remove this artifact [55–57].

Ponton and Eggermont described the LLAEP changes in amplitude and latency from infancy to adulthood as a measurement of auditory cortex maturation. In the case of normal hearing children, from 5 to 9 years old, the typical response has a large positive peak around 100-ms labeled P1 followed by a negative peak N1, P1–N1 complex; the amplitude and latency of this positive peak decrease as a function of age. This peak begins to spread out and finally divides into two positive peaks separated by a negative peak; from 10 years of age onwards, in this moment, the LLAEPs are similar to an adult morphology [58].

In this section, our research about the use of ICA not only to reduce CI artifact (spatial filtering) but also to detect LLAEPs (source extraction) is summarized. It explains how the parameter of different ICA algorithms was selected for optimal estimation of the LLAEP components. Additionally, a method for objectively selection of LLAEP components based on mutual information (MI) and clustering analysis is described; moreover, a procedure to identify ICs with *physiological meaning* as well as the ICs associated with the CI artifact is explained. Finally, we expose how denoising LLAEPs could be used for a robust evaluation of the maturation of the auditory system in children with CIs.

4.2   A Multichannel Dataset of Long Latency Auditory Evoked Potentials

Different types of EEG activity of different origin occurs simultaneously at diverse locations on the head, and so, the use of multiple electrodes for simultaneous recordings is encouraged. In addition, multichannel LLAEP recordings have been used to objectively study the maturation of the auditory system in young children; these recordings provide temporal resolution for the chronological aspects of brain plasticity [51].

The child-CI system is has been studied in the same way; however, as mentioned before, this is problematic as normal operation of the CI generates an electrical artifact; the CI artifact generally masks, either partially or totally, the brain auditory response and so results in errors in the analysis of the auditory response. Solving the CI artifact problem, the auditory neuroplasticity in children with CIs can be assessment by the changes of the LLAEP waveform attributed to the length of time of use/implantation of the CI on the child [59, 60].

In the following sections, we show the analysis of multichannel scalp EEG recordings from 10 deaf prelinguistic children from 6 to 9 years old. EEG were recorded using 19 electrodes placed according to the 10–20% electrode system, with reference to the linked mastoids. The EEG was sampled at 2 kHz and filtered online between 0.1 and 500 Hz. The recordings consisted of 150 epochs, 300 ms long (including 150 ms pre-stimulus in each epoch); automatic artifact rejection was used if the signal exceeded ±70 μV. The stimulus used to evoke the LLAEP consisted of a 50 ms tone burst (10-30-10) at 1 kHz and 70 dBHL.

4.3   Multichannel ICA

Most applications of ICA are to multichannel time series measurements. Regarding biomedical signals, different ICA algorithms in the literature have various statements about the number of sources and recording channels as well as nature of the noise or artifacts. The main assumptions made on applying multichannel or standard ICA to a measured signal such as EEG are summarized below:

  (i) The measured signals $x(t)$ are a result of a linear mixing of different sources; volume conduction in the brain results in linear and instantaneous mixing, and then, EEG recordings at the electrodes are assumed to be a linear mixture of the underlying brain sources and the artifact signals. ICA assumes that different physical process tend to generate different statistically independent signals [23].

 (ii) Another restriction in standard ICA is that the number of underlying sources is usually less than or equal to the number of measurements ($p \geq q$). The dataset used in our researches includes the recordings from 19 EEG electrodes; the numbers of stable estimations expected are variable across subjects, but in general, they could be one or two sources related to the auditory response, two to three linked to the CI artifact, and one or two other artifacts such as blinking and line noise; then, p will be a maximum of about six to seven independent components.

(iii) The sources are non-Gaussian and the measured signal is stationary (over the short epoch measured). The CI artifact happens at the same time in each EEG epoch and is time-locked with the stimuli; therefore, it is considered stationary and with a non-Gaussian power density function [57]. Furthermore, the ICA components can be estimated but with certain indeterminacies, for example, arbitrary scaling and permutation [22].

Figure 10 outlines the application of the multichannel ICA algorithm to our signal of interest. The electrical activity produced by different brain sources is recorded using the 19 electrodes of EEG. Although the LLAEP and CI are

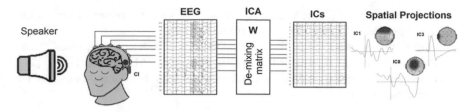

Fig. 10 CI receiver is listening to a sound (burst tone) by a speaker; EEG is recording using 19 scalp electrodes. ICA calculates the demixing matrix W used to estimate the brain sources and artifacts; the spatial projections of the independent components (ICs) are useful to identify the part of the scalp responsible for each estimate. (IC1, eye movements; IC3, CI artifact at T5; IC8, LLAEP with front-central distribution)

temporally correlated, they are spatially independent signals, since the CI artifact is generated by the device and not by a brain source [57]. Using the EEG, multichannel ICA calculates the mixing matrix A which depends on the conductivity characteristics of the brain and where the electrodes are placed; the demixing matrix used to estimate the sources is $W = A^{-1}$. Finally, ICA indicates what parts of the scalp are most responsible for the activity (auditory in our case) by interpolated topographic maps of the ICs. The columns of W^{-1} give the relative projection strengths of the respective components onto each of the scalp electrodes [53]. These topographic maps indicate the physiological origin of the estimated sources, for example, eye blinking and muscle activity.

4.3.1 ICA Algorithms: High-Order Statistic (HOS) Based Versus Second-Order Statistic (SOS) Based

Although a considerable amount of literature has been published on ICA algorithms, three algorithms can be classed as the most popular: *JADE* [30], *FastICA* [28], and *Infomax* [26]. These algorithms have been modified, improved, or extended by different authors; 30 ICA algorithms are included in the Matlab toolbox implemented by the group of Cichocki (ICALAB).[1]

One possible classification criterion of the different ICA algorithms could be the means of assessing independence; for example, if only the second-order statistics of the observations is used, the algorithm is called second-order statistic (SOS) ICA. Otherwise, it is called high-order statistics (HOS) ICA. Some general differences between the SOS and HOS are as follows: in SOS methods, the principal assumption is that the sources have some temporal structure, while the HOS methods minimize the MI, as a measurement of the independence of two variables, between the source estimates. The HOS methods cannot be applied to Gaussian signals, as the method does not allocate more than one Gaussian source. Additionally, the SOS

---

[1]Cichocki et al. [75].

methods do not permit the separation of sources with identical power spectra shape, independent, and identically distributed sources [61].

Bearing this in mind, the performance of two HOS ICA methods, FastICA and Infomax, and one SOS ICA algorithm, a modification of JADE called TDSep [27], were compared to determinate their optimal parameters for the estimation of the LLAEP components.

4.3.2 Optimal ICA Parameters for the Estimation of the LLAEP Components

Although AEP/LLAEP is one of the most used recordings to evaluate the performance of ICA algorithms in the literature and it has been demonstrated that this procedure can remove the typical EEG artifacts [9, 52, 62] and the specific artifact generated by the normal functioning of a CI [55, 57, 63], there are few studies about the selection of the optimal parameters for estimating the LLAEP components, to reliably recover both the auditory response and the CI artifact.

Using the method proposed by Himberg et al. to investigate the algorithmic and statistical reliability of the ICs recovered by FastICA [64], we select the model parameters more convenient for robust LLAEP and CI artifact estimation recovered by this algorithm. By running the ICA algorithm many times for random initial conditions and bootstrapping the data every time, the algorithmic and statistical reliability of the ICs recovered is investigated. The estimated components are clustered according to their mutual similarities; the criterion applied by Himberg et al. is an agglomerative clustering with average-linkage. Clusters are visualized as a 2D plot, and their robustness is measured by quality (stability) index, I_q (see Eq. 7).

A way to measure the similarity between the estimates is by the absolute value of their mutual correlation coefficients r_{ij}, $i, j = 1, 2, \ldots, k$; the final similarity matrix has the elements Δ_{ij} defined by $\Delta_{ij} = |r_{ij}|$. Then, I_q is calculated as follows:

$$I_q \left(C_{m_{\text{int}}} \right) = \frac{1}{\left| C_{m_{\text{int}}} \right|^2} \sum_{i, j \in C_{m_{\text{int}}}} \Delta_{ij} - \frac{1}{\left| C_{m_{\text{int}}} \right| \left| C_{m_{\text{ext}}} \right|} \sum_{i \in C_{m_{\text{int}}}} \sum_{j \in C_{m_{\text{int}}}} \Delta_{ij} \qquad (7)$$

If C_m denotes the set of indices of all the estimates, $C_{m_{\text{int}}}$ the set of indices that belong in the m-th cluster, and $\left| C_{m_{\text{int}}} \right|$ the size of the m-th cluster, then I_q is computed as the difference between the average intra-cluster similarities and the average extra-cluster similarities; $C_{m_{\text{ext}}}$ is the set of indices that do not belong to the m-th cluster. The cluster quality index gives a rank of the corresponding IC clusters estimated. The ideal value of I_q is 1; the smaller the value, the less stable, compact, and isolated the estimated cluster is.

In the case of FastICA, three nonlinear functions and two orthogonalization approaches were compared using I_q. The nonlinear functions $G_1(y) = y^3$,

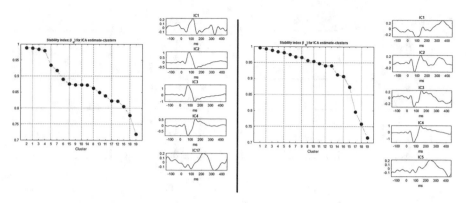

Fig. 11 Comparison using I_q between test conditions $G_2(y) = \tan h(y)$ with deflation (left) and $G_1(y) = y^3$ with symmetric approach in a recording from a child with CI. The second test condition achieved the most robust clusters as well as the most reliable estimates for both LLAEP and CI artifact

$G_2(y) = \tan h(a_1 y)$, and $G_3(y) = y \exp\left[-a_2 \frac{y^2}{2}\right]$ are used together with deflationary and symmetric orthogonalization approaches. The default value for a_1 and a_2 is 1.

Figure 11 shows a comparison using I_q between test conditions $G_2(y) = \tan h(y)$ with deflation (left) and $G_1(y) = y^3$ with symmetric approach in a recording from a child with CI. For the first test condition, only six estimated clusters have I_q between 0.9 and 1, and none of those clusters are related to the LLAEP, only with the CI artifact; IC17 with $I_q = 0.82$ is related to the auditory response. On the other hand, the second condition has the most robust clusters with I_q between 0.9 and 1; the clusters ranked first (cluster 1 and 4) correspond to CI artifact, while cluster 5 (with a I_q index value lower than 0.9) is associated with the LLAEP.

For FastICA, the highest values of the I_q index, in other word the most stable clusters, were obtained using the nonlinear function $G_1(y)$ together with a *symmetric* orthogonalization approach. The estimate clusters ranked first (the highest I_q index values) were generally related to the CI artifact and the LLAEP. Additionally, the number of clusters with a stability index between 0.9 and 1 was greater using this test condition [56, 65].

In order to compare the estimated components recovered using Infomax and Ext-Infomax, the kurtosis values and the probability density function (pdf) of the estimates were used; Also, if it was possible, the waveforms of the estimates were associated with the components of interest, i.e., LLAEP, CI artifacts, and noise. Figure 12 shows pdfs and kurtosis values for selected estimates for one of the recordings analyzed using Infomax and Ext-Infomax.

Kurtosis values of the estimates in both original Infomax and Ext-Infomax for the LLAEPs and CI artifact are similar; the values for the estimates of the LLAEPs are positive, as expected [66]. The kurtosis values of the CI artifact estimate depend on the part of the artifact recovered (the transient at the beginning and/or end of the

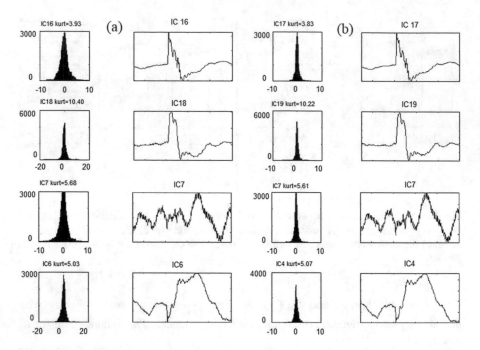

Fig. 12 Estimates obtained from a recording of a child who has used a cochlear implant for 2.5 year; there are no differences neither in the pdf histogram nor in the kurtosis values between (**a**) Infomax and (**b**) Ext-Infomax. The LLAEPs can be recognized in the ICs of the rows 3 and 4; however, the estimates are contaminated by the background noise or the CI artifact

artifact or the stimuli pulses). The principal differences between Infomax and Ext-Infomax are in the noise estimate components; Ext-Infomax is more appropriate to estimate sources with pdfs close to Gaussian distributions. Based on the kurtosis values, Ext-Infomax was finally selected since the noise recovered is better than in the original Infomax, which result in an easier identification of the estimates related to the LLAEPs [56, 65].

TDSep is based on several time-delayed correlation matrices (τ); the τ parameter must be chosen to take advantage of the temporal structure of the signals. Meinecke et al. [31] propose the use of resampling methods to assess the reliability of this algorithm and the variance of the estimates as a measure of the separation error (separability matrix); a low value corresponds to a good separation. To determinate the best time delay parameter for the dataset analyzed, τ for TDSep was varied from 1 to 20 in steps of 5. It was selected such that τ had the lowest separability matrix values and a clear block structure in the matrix, where the LLAEPs, CI artifact, and background noise could be recognized.

Figure 13 shows a comparison between the separability matrix using a time delay $\tau = 0, \ldots, 1$ (left) and $\tau = 0, 1, 2, \ldots, 20$ (right) for one recording from a child with CIs. The separability values for the LLAEP estimated are smaller for $\tau = 0, 1, 2, \ldots, 20$ than $\tau = 0, \ldots, 1$. Using a time delay from 0 to 1, only the CI artifact can

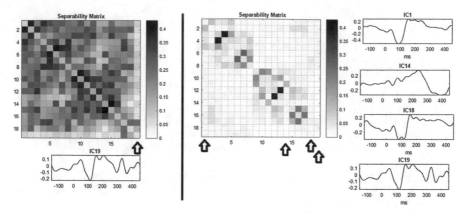

Fig. 13 Comparison between the separability matrixes with two different time delays using TDSep, left column $\tau = 0, \ldots, 1$ and right column $\tau = 0, 1, 2, \ldots, 20,$. In general, the separability matrix values were the lowest with $\tau = 0, 1, 2, \ldots, 20$; in most of the recordings, one ICs can be associated with the LLAEP and CI artifact using this time delay

be recovered clearly (IC19), while using a $\tau = 0, 1, 2, \ldots, 20$, more components are related to the CI artifact (IC1, IC18, and IC19) and one component can be associated with the LLAEP (IC14), although the CI artifact is still being mixed with IC14 (neither FastICA nor Ext-Infomax recovered the LLAEP in this recording).

Clearer block structures were identified using TDSep with $\tau = 0, 1, 2, \ldots, 20$ than with $\tau = 0, \ldots, 1$, in all recordings analyzed, resulting in a notable separation between the ICs related to the CI artifact and the LLAEP estimates [56, 66].

4.3.3 Objective IC Selection of AEPs in Children with Cochlear Implants Using Clustering

Krashov et al. [67] propose to use MI as a similarity measure for hierarchical clustering of the ICs computed by ICA from the ECG of pregnant women; MI values between variables satisfy the conditions to cluster objects (positive, symmetric, and equal to zero only if the variables are the same or the variables are independent). This method was slightly modified in this part of our research, in order to objectively select the ICs associated with the LLAEP, from ongoing EEG recorded from children with CIs, as well as to identify the ICs related to the CI artifact. The procedure introduced here to objectively select ICs has two steps (see Fig. 14), the first to reduce the number of estimates to compute and the second to cluster the most robust estimates [68]. The two steps which summarize our procedure are explained in the following paragraphs.

Step 1: Reduction of number of electrodes: If we consider that more channels of measurements generally imply more complex calculations, reducing the number of channels included in the estimation might help reduce the complexity. We

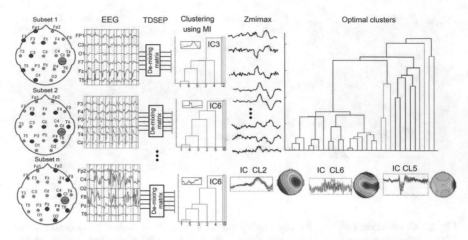

Fig. 14 Outline of the procedure to objectively select consistent ICs through MI and clustering. For each measurement subset (six electrodes in this case), the ICs were calculated using TDSep; these were then grouped together in accordance with the values of MI between them, and with a dendrogram using MI, the IC last merged was selected, in this case, IC3 for subset 1, IC6 for the subset n, etc.; Z_{mimax} is a matrix with ICs with maximum MI. The ICs selected in the previous step were clustered using the Euclidean distance and average-linked hierarchical clustering; the final optimal number of clusters (CL) was determined using the procedure explained in Sect. 4.4.1

propose to randomly select a subset of six channels, from all the channels, making sure that, in each selection, channels representative of all the areas of the brain are included (frontal, central, temporal, etc.). Using TDSep with a time delay $\tau = 0, 1, 2, \ldots, 20$, the ICs for each subset of electrodes obtained pseudo-randomly in the reduction of electrodes were calculated. The MI between the ICs in each subset is calculated and used as a similarity measure to the cluster analysis. Using the residual MI between the ICs, the estimate with the minimal dependency with the rest of the estimations is selected in each subset; the IC last merged was selected in each subset with the purpose of forming a new matrix data (Z_{mimax}). The dendrogram construction was reformulated by finding the distance between the elements of the new matrix; the proximity of the ICs is defined as the minimum distance (maximum of the MI). The minimally dependent IC in subset 1 corresponds to the AEP, while it corresponds to artifacts in subsets 2 and n.

Step 2: Optimal clustering estimates: Different agglomerative methods, using four similarity functions (Euclidean, Cosine, Spearman, and City Block) and three link criteria (average, complete, and single), were compared using Cophenetic Correlation Coefficient (CCC); three criteria for cluster validation, Calinski-Harabasz, Davies-Bouldin, and Silhouette, were used to determine optimal clustering LLAEP ICs (see Sect. 4.4.1 for details). Only the clusters with more than one estimate were considered as robust clusters (IC CL3, IC CL5, and IC CL6 in this example). The trace in red is the average of the estimation in each cluster.

4.4 Results

4.4.1 Optimal Clustering LLAEPs in Subjects with CI

After forming the Z_{mimax} (see Fig. 14), four distance measures were compared to calculate the proximities between the data, Euclidean, Cosine, Spearman, and City Block. Additionally, three different agglomerative methods were used to cluster the Z_{mimax}: (1) Single-link clustering defines the distance between groups as that of the closest pair of individuals. (2) Complete-linking clustering is the opposite of single-link clustering and defines distance between groups as that of the most distant pair of individuals. (3) Average-link clustering defines distance between groups as the average of the distances between all pairs of individuals [69].

The CCC was used to compare the quality cluster solutions obtained by the different combinations of the metrics mentioned before. Finally, three inter-/intra-quality indices were used to find the optimal number of clusters: Index 1, Calinski-Harabasz; Index 2, Davies Bouldin; and Index 3, Silhouette [69].

The performance of both the Euclidean and City Block function is high and comparable in the three link-clustering modalities: average, complete. and single. In general, Cosine and Spearman performance, also in the three modalities mentioned before, is lower than the previous two functions. The three indices of inter-/intra-cluster evaluation show very similar values to optimal cluster number, $k\sim8$. Finally, it is possible to say that using the Euclidean similarity function with link-clustering average, robust clusters were obtained, and it was also possible to identify the auditory response in those clusters [70].

Using the scalp map of each cluster obtained with the procedure explained in the previous paragraphs, it was possible to identify the spatial distribution of relevant ICs. It is possible to say that this procedure could help to (1) know the number of sources to be estimated by ICA, (2) recognize the ICs related with the auditory response and/or a specific EEG artifact, and (3) identify the spatial distribution of each cluster related to LLAEPs.

4.4.2 Maturation of the Auditory System in Children with Cochlear Implants

After removing the CI artifact of EEG, it was possible to study the maturation of the auditory system in these children. Comparing LLAEP recordings from children using their CIs for less than 1 year (5 months on average) and recordings of the same subjects a year after their implantation, it is possible to observe the following: (1) Changes in the latency of the P_1 peak of the ICs associated with the LLAEP between both recordings varied among subjects. (2) Changes in the spatial projections are similar between subjects, changing from parietal (first recording) to front-central at 1 year of implant use.

Finally, it was possible to study the interhemispheric asymmetry (IA) in LLAEPs; IA is a measure of the cortical reorganization associated with profound unilateral deafness [71]. Ponton et al. analyzed the IA, due to changes in the cortical activation in teens and adults with late-onset profound unilateral deafness, using the LLAEP peak N_1 amplitude cross-correlation coefficient. The authors suggest that the IA may result from gradual changes in the brain generators of N_1 and that these changes continue at least 2 years after the onset of hearing loss. In the case of children, Sharma et al. [72] used the latency of the P_1 wave of LLAEPs as a biomarker of the development and plasticity of the central auditory system in children with a hearing aid and/or CI receiver. The intraclass correlation coefficient (ICC) between mean global field powers (MGFPs) of right and left electrodes is proposed to quantify LLAEP IA as a measure of auditory cortex reorganization in CI recipients [73, 74]. The increases in amplitude and shape asymmetries of the LLAEP topographic map (visually observed) were reflected in a considerable reduction of ICC values (on average 41.4%), at more than 2 years postimplantation surgery.

4.5   Discussion and Conclusions

The optimal parameters to recover both the LLAEP and the CI artifact for the three ICA algorithms assessed in sections are as follows: (a) FastICA with a *symmetric* orthogonal approach and the nonlinear function $G_1(y) = y^3$, (b) Ext-Infomax instead of Infomax, and (c) TDSep with time delay $\tau = 0, 1, 2, \ldots, 20$.

FastICA with this test condition achieved the highest number of clusters with I_q index values between 0.9 and 1 (in 9 of 10 recordings analyzed); the robustness of clusters and reliability of the estimates were better than any other test condition. Most of the times, those clusters are related to the CI artifact components and noise but are not necessarily related to the LLAEP. The principal differences between the estimated components using Infomax and Ext-Infomax were in the background noise; the kurtosis values of the estimates change from positive (close to zero) to negative; the pdf histograms have different shapes for the LLAEP, CI artifact, and noise using Ext-Infomax. Finally, TDSep with $\tau = 0, 1, 2, \ldots, 20$ achieved the lowest separability matrix values, and the structure of the matrix is the clearest over all the time delays evaluated.

The numbers of clusters and the similarity function that yield best results in these datasets, in other words optimal clustering LLAEP ICs, were 8 and Euclidean link-clustering average, respectively.

In conclusion, we have applied multichannel ICA not only to reduce the CI artifact but also to detect the LLAEPs and to use the changes in the spatial projections of their ICs for a robust evaluation of the maturation of the auditory system in children with CIs [74]. Our studies of the last 10 years offer a useful tool to evaluate cortical reorganization after CI implantation by comparing recordings longitudinally. The LLAEP analysis summarized in this section could be useful to make decisions about changes in the CI settings or in the rehabilitation strategy of children with CIs.

5 Final Remarks

ICA is a powerful signal processing tool that, while providing promising solutions on the separation of multiple and statistically independent components from a set of mixtures, brings a challenge in the field of biomedical signal processing, automatic identification, and classification of meaningful components. In this chapter, the classification of such components has been either (1) focused on rhythmic physiological events such as the fetal heart sounds or (2) focused on synchronized and proximal (spatially localized) events like the cochlear implant artifact. In both case studies, the classification stage has been fundamental on the performance of the denoising process, on the quality of the retrieved signals, and, finally, on any further analysis that may be applied on them for the recovery of physiological information of interest.

References

1. Hyvärinen, A. (2013). Independent component analysis: Recent advances subject areas. *Philosophical Transactions of the Royal Society A, 371*(20110534), 1–19.
2. Stone, J. V. (2004). *Independent component analysis: A tutorial introduction.* Cambridge, MA: MIT Press.
3. Hyvärinen, A., & Oja, E. (2000). Independent component analysis: Algorithms and applications. *Neural Networks, 13*(4–5), 411–430.
4. Vincent, E., Araki, S., Theis, F., & Nolte, G. (2012). The signal separation evaluation campaign (2007–2010): Achievements and remaining challenges. *Signal Processing, 92*(8), 1928–1936.
5. James, C. J., & Hesse, C. W. (2005). Independent component analysis for biomedical signals. *Physiological Measurement, 26*(1), R15–R39.
6. Tharwat, A. (2018). Independent component analysis: An introduction. *Applied Computing Informatics, In Press*, 1–15.
7. Hyvärinen, A. (1999). Survey on independent component analysis. *Neural Computation Survey, 2*, 94–128.
8. Ejaz, M. (2008). *A framework for implementing independent component analysis algorithms.* Department of Electrical & Computer Engineering, Florida State University.
9. Jung, T., Makeig, S., Lee, T., Mckeown, M. J., Brown, G., Bell, A. J., & Sejnowski, T. J. (2000). Independent component analysis of biomedical signals. In: *2nd international workshop on independent component analysis and blind signal separation, no. 1* (pp. 633–644).
10. Klemm, M., Haueisen, J., & Ivanova, G. (2009). Independent component analysis: Comparison of algorithms for the investigation of surface electrical brain activity. *Medical & Biological Engineering & Computing, 47*(4), 413–423.
11. Kuzilek, J. (2013). *Independent component analysis: Applications in ECG signal processing.* Prague: Department of Cybernetics, Czech Technical University.

12. Staudenmann, D., & Daffertshofer, A. (2007). Independent component analysis of high-density electromyography in muscle force estimation. *IEEE Transactions on Biomedical Engineering, 54*(4), 751–754.

13. Hild, K. E., Alleva, G., Nagarajan, S., & Comani, S. (2007). Performance comparison of six independent components analysis algorithms for fetal signal extraction from real fMCG data. *Physics in Medicine and Biology, 52*(2), 449–462.

14. Comani, S., & Alleva, G. (2007). Fetal cardiac time intervals estimated on fetal magnetocardiograms: Single cycle analysis versus average beat inspection. *Physiological Measurement, 28*(1), 49–60.

15. Cao, J., Murata, N., Amari, S., Cichocki, A., & Takeda, T. (2003). A robust approach to independent component analysis of signals with high-level noise measurements. *IEEE Transactions on Neural Networks, 14*(3), 631–645.

16. Wübbeler, G., Ziehe, A., Mackert, B. M., Müller, K. R., Trahms, L., & Curio, G. (2000). Independent component analysis of noninvasively recorded cortical magnetic DC-fields in humans. *IEEE Transactions on Biomedical Engineering, 47*(5), 594–599.

17. Jiménez-González, A., & James, C. J. (2009). Extracting sources from noisy abdominal phonograms: A single-channel blind source separation method. *Medical & Biological Engineering & Computing, 47*(6), 655–664.

18. Jimenez-Gonzalez, A., & James, C. J. (2008). Blind source separation to extract foetal heart sounds from noisy abdominal phonograms: A single channel method. In: *4th IET international conference advances medical, signal information processing (MEDSIP 2008)* (pp. 114–118).

19. Jimenez-Gonzalez, A., & James, C. (2008). Source separation of foetal heart sounds and maternal activity from single-channel phonograms: A temporal independent component analysis approach. *Computers in Cardiology, 2008*(Md), 949–952.

20. Jiménez-González, A., & James, C. J. (2013). Blind separation of multiple physiological sources from a single-channel recording: A preprocessing approach for antenatal surveillances. *IX International Seminar on Medical Information Processing and Analysis, 8922*, 1–11.

21. Comon, P. (1994). Independent component analysis, a new concept? *Signal Processing, 36*, 287–314.

22. Cardoso, J.-F. (2009). Blind signal separation: Statistical principles. *Proceedings of the IEEE, 86*(10), 2009–2025.

23. Amari, S., Cichocki, A., & Yang, H. H. (1989). *A new learning algorithm for blind signal separation.* San Mateo: Morgan Kaufmann Publishers.

24. Hyvärinen, A., & Oja, E. (1997). A fast fixed-point algorithm for independent Componet analysis. *Neural Computation, 9*, 1483–1942.

25. Bell, A. J., & Sejnowski, T. J. (1995). An information-maximisation approach to blind separation and blind deconvolution. *Neural Computation, 7*(6), 1129–1159.

26. Lee, T.-W., Girolami, M., & Sejnowski, T. J. (1999). Independent component analysis using an extended Infomax algorithm for mixed Subgaussian and Supergaussian sources. *Neural Computation, 11*(2), 417–441.

27. Ziehe, A., & Müller, K.-R. (1998). TDSEP – an efficient algorithm for blind separation using time structure. In L. Niklassion, M. Boden, & G. Ziemke (Eds.), *Proceedings international conference artificial neural networks* (pp. 675–680). Berlin: Springer.

28. Hyvärinen, A. (1999). Fast and robust fixed-point algorithms for independent component analysis. *IEEE Transactions on Neural Networks, 10*(3), 626–634.

29. Hyvärinen, A., & Oja, E. (2000). Independent component analysis: Algorithms and applications. *Neural Networks, 13*(4–5), 411–430.

30. Cardoso, J., & Souloumiac, A. (1993). Blind beanforming for non -Gaussian signals. *IEE Proceedings F-Radar and Signal Processing, 140*(6), 362–370.

31. Meinecke, F., Ziehe, A., Kawanabe, M., & Muller, K. R. (2002). Resampling approach to estimate the stability of one-dimensional or multidimensional independent components. *IEEE Transactions on Biomedical Engineering, 49*(12), 1514–1524.

32. Davies, M., & James, C. (2007). Source separation using single channel ICA. *Signal Processing, 87*(8), 1819–1832.

33. Acharyya, R., Scott, N. L., & Teal, P. (2009). Non-invasive foetal heartbeat rate extraction from an underdetermined single signal. *Health (Irvine, California), 01*(02), 111–116.
34. Kovács, F., Horváth, C., Balogh, A. T., & Hosszú, G. (2011). Fetal phonocardiography–past and future possibilities. *Computer Methods and Programs in Biomedicine, 104*(1), 19–25.
35. Chourasia, V. S., Tiwari, A. K., & Gangopadhyay, R. (2014). A novel approach for phono-cardiographic signals processing to make possible fetal heart rate evaluations. *Digital Signal Processing, 30,* 165.
36. Várady, P., Wildt, L., Benyó, Z., & Hein, A. (2003). An advanced method in fetal phonocar-diography. *Computer Methods and Programs in Biomedicine, 71*(3), 283–296.
37. Ruffo, M., Cesarelli, M., Romano, M., Bifulco, P., & Fratini, a. (2010). An algorithm for FHR estimation from foetal phonocardiographic signals. *Biomedical Signal Processing and Control, 5*(2), 131–141.
38. Mittra, A. K., & Choudhari, N. K. (2009). Development of a low cost fetal heart sound monitoring system for home care application. *Journal of Biomedical Science and Engineering, 2*(6), 380–389.
39. Jimenez-Gonzalez, A. (2010). *Antenatal foetal monitoring through abdominal phonogram recordings: A single-channel independent component analysis approach.* Institute of Sound and Vibration Research, University of Southampton.
40. Broomhead, D. S., & King, G. P. (1986). Extracting qualitative dynamics from experimental data. *Physica D: Nonlinear Phenomena, 20*(2–3), 217–236.
41. Holburn, D., & Rowsell, T. (1989). Real time analysis of fetal phonography signals using the TMS320. In: *Biomedical applications of digital signal processing, IEE colloquium on, 1989* (pp. 7–1).
42. Talbert, D., Davies, W. L., Johnson, F., Abraham, N., Colley, N., & Southall, D. P. (1986). Wide Bandwidlt fetal phonography using a sensor matched to the compliance of the Mother's Abdominal Wall. *Biomedical Engineering IEEE Transactions, BME-33*(2), 175–181.
43. Hyvärinen, A., & Oja, E. (1997). A fast fixed-point algorithm for independent component analysis. *Neural Computation, 9*(7), 1483–1492.
44. Jiménez-González, A., & James, C. J. (2010). Time-structure based reconstruction of physio-logical independent sources extracted from noisy abdominal phonograms. *IEEE Transactions on Biomedical Engineering, 57*(9), 2322–2330.
45. Jiménez-González, A., & James, C. J. (2012). On the interpretation of the independent components underlying the abdominal phonogram: A study of their physiological relevance. *Physiological Measurement, 33*(2), 297–314.
46. Golyandina, N., Nekrutkin, V., & Zhigljavsky, A. A. (2001). *Analysis of time series structure: SSA and related techniques.* Boca Raton: Chapman & Hall/CRC.
47. Guijarro-Berdiñas, B., Alonso-Betanzos, A., & Fontenla-Romero, O. (2002). Intelligent analysis and pattern recognition in cardiotocographic signals using a tightly coupled hybrid system. *Artificial Intelligence, 136*(1), 1–27.
48. Zuckerwar, A. J., Pretlow, R. A., Stoughton, J. W., & Baker, D. A. (1993). Development of a piezopolymer pressure sensor for a portable fetal heart rate monitor. *Biomedical Engineering IEEE Transactions, 40*(9), 963–969.
49. Ruffo, M., Cesarelli, M., Romano, M., Bifulco, P., & Fratini, A. (2010). A simulating software of fetal phonocardiographic signals. In: *Information technology and applications in biomedicine (ITAB), 2010 10th IEEE international conference on* (pp. 1–4).
50. Jiménez-González, A., & James, C. J. (2013). Antenatal surveillance through estimates of the sources underlying the abdominal phonogram: A preliminary study. *Physiological Measurement, 1041,* 1041–1061.
51. Picton, T. W., Bentin, S., Berg, P., Donchin, E., Hillyard, S. a., Johnson, R., Miller, G. a., Ritter, W., Ruchkin, D. S., Rugg, M. D., & Taylor, M. J. (2000). Guidelines for using human event-related potentials to study cognition: Recording standards and publication criteria. *Psychophysiology, 37*(2), 127–152.

52. Jung, T., Humphries, C., Lee, T., Makeig, S., Mckeown, M. J., Iragui, V., & Sejnowski, T. J. (1998). Extended ICA removes artifacts from electroencephalographic recordings. *Advances in Neural Information Processing Systems, 10*, 894–900.
53. Delorme, A., & Makeig, S. (2004). EEGLAB: An open source toolbox for analysis of single-trial EEG dynamics including independent component analysis. *Journal of Neuroscience Methods, 134*(1), 9–21.
54. Delorme, A., Sejnowski, T., & Makeig, S. (2007). Enhanced detection of artifacts in EEG data using higher-order statistics and independent component analysis. *NeuroImage, 34*(4), 1443–1449.
55. Campos Viola, F., Thorne, J., Edmonds, B., Schneider, T., Eichele, T., & Debener, S. (2009). Semi-automatic identification of independent components representing EEG artifact. *Clinical Neurophysiology, 120*(5), 868–877.
56. Castañeda-Villa, N., & James, C. J. (2011). Independent component analysis for auditory evoked potentials and cochlear implant artifact estimation. *IEEE Transactions on Biomedical Engineering, 58*(2), 348–354.
57. Gilley, P. M., Sharma, A., Dorman, M., Finley, C. C., Panch, A. S., & Martin, K. (2006). Minimization of cochlear implant stimulus artifact in cortical auditory evoked potentials. *Clinical Neurophysiology, 117*, 1772–1782.
58. Ponton, C. W., Don, M., Eggermont, J. J., Waring, M. D., & Masuda, A. (1996). Maturation of human cortical auditory function: Differences between normal-hearing children and children with cochlear implants. *Ear and Hearing, 17*, 430–437.
59. Gilley, P. M., Sharma, A., & Dorman, M. F. (2008). Cortical reorganization in children with cochlear implants. *Brain Research, 1239*(1999), 56–65.
60. Sharma, A., Gilley, P. M., Dorman, M. F., & Baldwin, R. (2007). Deprivation-induced cortical reorganization in children with cochlear implants. *International Journal of Audiology, 46*, 494–500.
61. Cichocki, A., & Amari, S. (2002). Robust techniques for BSS and ICA with Noisy data. In *Adaptive blind signal and image processing* (pp. 307–308). West Sussex: Wiley.
62. Vigário, R. N. (1997). Extraction of ocular artefacts from EEG using independent component analysis. *Electroencephalography and Clinical Neurophysiology, 103*, 395–404.
63. James, C., & Castañeda-Villa, N. (2006). ICA of auditory evoked potentials in children with cochlear implants: Component selection. In: *IET 3rd international conference on advances in medical, signal and information processing, 2006. MEDSIP 2006* (pp. 6–9).
64. Himberg, J., & Hyvarinen, A. (2003). Icasso: Software for investigating the reliability of ICA estimates by clustering and visualization. 2003 *IEEE XIII Work. Neural Networks Signal Process (IEEE Cat. No.03TH8718)* (pp. 259–268).
65. Castañeda-Villa, N., & James, C. J. (2008). The selection of optimal ICA component estimates using 3 popular ICA algorithms. In: *Annual international IEEE EMBS conference* (pp. 5216–5219).
66. Jung, T.-P., Humphriesl, C., Lee, T., Makeig, S., Mckeown, M. J., Iragui, V., & Sejnowski, T. J. (1998). Extended ICA removes artifacts from electroencephalographic recordings. *Advances in Neural Information Processing Systems, 10*, 894–900.
67. Kraskov, A., Stögbauer, H., Andrzejak, R. G., & Grassberger, P. (2005). Hierarchical clustering using mutual information. *Europhys, 70*(2), 278–284.
68. Castaneda-Villa, N., & James, C. J. (2007). Objective source selection in blind source separation of AEPs in children with Cochlear implants. In: *Proceedings 29th annual international conference IEEE EMBS* (pp. 6223–6226).
69. Everitt, B., & Hothorn, T. (2011). *An introduction to applied multivariare analysis with R*. New York: Springer.
70. Castañeda-Villa, N., Cornejo-Cruz, J. M., & Granados-Trejo, P. (2015). Comparison between different similarity measure functions for optimal clustering AEPs independent components. *IEEE Engineering in Medicine and Biology Society, 2015*, 7446–7449.

71. Ponton, C. W., Vasama, J.-P., Tremblay, K. L., Khosla, D., Kwong, B., & Don, M. (2001). Plasticity in the adult human central auditory system: Evidence from late-onset profound unilateral deafness. *Hearing Research, 154*, 32–44.
72. Sharma, A., Martin, K., Roland, P., Bauer, P., Sweeney, M. H., Gilley, P. M., & Dorman, M. (2005). P1 latency as a biomarker for central auditory development in children with hearing impairment. *Journal of the American Academy of Audiology, 16*(8), 564–573.
73. Castañeda-Villa, N., Cornejo-Cruz, J. M., & James, C. (2009). Independent component analysis for robust assessment of auditory system maturation in children with cochlear implants. *Cochlear Implants International, 11*(2), 71–83.
74. Castañeda-Villa, N., Cornejo, J. M., James, C. J., & Maurits, N. M. (2012). Quantification of LLAEP interhemispheric symmetry by the intraclass correlation coefficient as a measure of cortical reorganization after cochlear implantation. *International Journal of Pediatric Otorhinolaryngology, 76*(12), 1729–1736.
75. Cichocki, A., Amari, S., Siwek, K., Tanaka, T., Anh Huy Phan, et al. *ICALAB toolboxes*. http://www.bsp.brain.riken.jp/ICALAB

Pattern Recognition Applied to the Analysis of Genomic Data and Its Association to Diseases

Verónica Jiménez-Jacinto, Laura Gómez-Romero,
and Carlos-Francisco Méndez-Cruz

Abstract The analysis of genomic data has been used to generate information about genetic variants and expression patterns correlated with specific physical traits. In the last decades, these analyses have evolved toward analyzing thousands of entities at the same time. Moreover, these analyses have produced an enormous amount of biomedical literature reporting associations between genes and diseases. In this scenario, pattern recognition techniques have been truly useful, so a review of how these techniques have been applied is relevant. Thus, in this chapter we present a brief introduction to the high-throughput sequencing methodologies. Then, we describe the process of identification of genomic variants and genetic expression profiles that have been used for the diagnostic of diseases, followed by a general overview of the gene-disease association extraction from biomedical literature.

Keywords Next-generation sequencing technologies · Genomic variation · Variant calling · Gene differential expression · Gene-disease association extraction · Information extraction · Biomedical natural language processing

V. Jiménez-Jacinto (✉)
Instituto de Biotecnologia de la Universidad Nacional Autónoma de México,
Cuernavaca Morelos, C.P. 62210, México
e-mail: vjimenez@ibt.unam.mx

L. Gómez-Romero
Instituto Nacional de Medicina Genómica, Ciudad de México, CDMX, México
e-mail: lgomez@inmegen.gob.mx

C.-F. Méndez-Cruz
Centro de Ciencias Genómicas de la Universidad Nacional Autónoma de México, Cuernavaca,
Morelos, México
e-mail: cmendezc@ccg.unam.mx

© Springer Nature Switzerland AG 2020
M. R. Ortiz-Posadas (ed.), *Pattern Recognition Techniques Applied to Biomedical Problems*, STEAM-H: Science, Technology, Engineering, Agriculture, Mathematics & Health, https://doi.org/10.1007/978-3-030-38021-2_2

35

1   Relevance of Applying Pattern Recognition Techniques to the Analysis of Genomic Data

Since the ending of the last century, the incorporation of methodologies for producing large-scale genomic information (microarrays, massive sequencing, etc.) has changed the paradigm of information analysis. While in the former paradigm, the analysis was focused on a single gene, its sequence, or its expression under diverse conditions or different tissues, the new technologies allow the generation of global information regarding the complete genome sequences, deriving into the possibility of obtaining a unique study of the global expression of all the genes within an organism.

Nowadays, the problem is not focused on generating information but on how to veridically analyze it. The information volumes have left behind the megabytes just to be expressed now in terms of gigabyte, terabytes, and even petabytes. The search for genomic variables, gene expression profiles linked to diseases, or the automatic analysis of biomedical literature requires the use of computing tools as well as statistics and pattern recognition techniques.

The specialists of areas such as Biology, Biochemistry, Veterinary and Health Sciences, which gather genomic information, have much more available information than in any previous times, but they face the problems of handling, summarizing, analyzing, and drawing feasible conclusions based on the large amount of generated information. For example, in recent years, vast amounts of gene expression data have been collected, which is surpassing the capacity of analysis. Much of that information is available in public databases, such as the Gene Expression Omnibus (GEO, https://www.ncbi.nlm.nih.gov/geo/).

This fact forces them not only to use computing tools to handle the data but also to apply statistical analysis strategies and pattern recognition techniques, which enable them to collect and classify information to obtain new knowledge. More than in any other times, the use of clustering techniques and automatic text analysis are being highly demanded.

The objective of this chapter is to offer an introduction to the massive sequencing data generation methodologies and the identification process of genetic variables to make genotype-phenotype associations. Then, the processes for generating genic expression patterns will be reviewed in order to further apply them in disease diagnosis or therapy selection using clustering processes. Finally, the application of pattern recognition for automatic analysis of biomedical literature to extract gene-disease associations will be surveyed.

2 Introduction to Genomic Analysis by High-Throughput Sequencing Technologies

This section is devoted to include a brief introduction to the experimental technologies for generating the raw data used in genomic analysis.

2.1 Genomes and Genes

The genome is the molecular structure in which all the information required to build an organism is encoded. It is composed of two strains, each one constituted of a basic alphabet of four nitrogenous bases (adenine, thymine, cytosine, and guanine) which are joined to a deoxyribose sugar and a phosphate molecule. The human genome contains roughly 3,000 million base pairs, and each cell of the human body contains one whole genome inside its nucleus.

The genome contains genes, which are stretches of DNA that encode for proteins or functional RNAs. According to the Human Genome Project, there are roughly 20,000 protein-coding genes in the human genome, which are translated into around 25,000 proteins [1].

The specific sequence of an organism genome and the level of expression of its genes have been found to associate with specific traits and diseases. So, the techniques used to decode the composition of the genome and the amount of gene molecules produced are of utmost importance in the genomic sciences. Microarrays and sequencing are the preferred methods to answer these two elemental questions.

2.2 Microarrays

The chemical structure of DNA allows the formation of hydrogen bonds between complementary nucleotide base pairs generating sequence-dependent hybridization between DNA strands. This property is exploited by microarrays technology [2].

In the case of genomic polymorphisms or variants, different probes which can either contain or not the variants of interest are printed into the microarray surface, a specialized machine is used to take pictures of the hybridization process, and the images are translated into the specific genotypes of each of the interrogated genomic positions. By other hand, expression microarrays contain sequences of genes of interest printed onto the surface. Labeled complementary DNA (cDNA) is hybridized against the microarray surface allowing the quantification of gene expression. The cDNA is produced from the RNA molecules present in the cells of interest, and these RNA molecules constitute the set of genes that are active [3].

Microarray technologies enable the interrogation of thousands of genes at the same time, allowing researchers to speed the process of data generation. However, the sequence of the probes must be designed and, as a consequence, only genomic regions previously described as polymorphic or as transcriptionally active can be interrogated. To overcome this caveat sequencing technologies can be used.

2.3 DNA-RNA Sequencing

Sequencing technologies permit the interrogation of the whole genome (DNA sequencing) or transcriptome (RNA sequencing) in one single experiment. The first technique used to sequence a whole genome was Sanger sequencing which was used to sequence the PhiX virus and then was applied by the Human Genome Project to generate the first draft of the human genome in 2001 [4]. Even though each step of the process was highly improved during the project, there were some doubts about its applicability for massive sequencing due to its low throughput and high cost [5]. However, massive parallel techniques have been created, and as a consequence, the cost has been abruptly diminished, and the throughput has been exponentially increased. As a consequence, every time is more common to do sequencing to identify the whole set of genomic variants of one individual or to quantify the expression level of the whole transcriptome of a cell population [6, 7]. Currently, there are a lot of sequencing alternatives which are based in a broad variety of chemical principles and experimental designs. The most common options are described in the following sections.

2.3.1 Sequencing by Ligation

This method is based on the ligation between a probe and the template of DNA. The probe is composed by one or two known bases and a stretch of degenerated bases. A fluorophore is released when the probe and the DNA template are ligated. In this way, the next base of the DNA that is being sequenced is identified. SOLiD and Complete Genomics applied this technology in their sequencing service.

2.3.2 Sequencing by Synthesis

The method of sequencing by synthesis can be further classified in sequencing by synthesis using reversible terminators and sequencing by synthesis by the addition of one single nucleotide.

The technology of reversible terminators uses labeled bases. Each one of the four possible bases is labeled with a different dye, and the 3'-end is blocked. Each sequencing cycle the four bases are added, the base that is complementary to the next base in the DNA template is incorporated by the polymerase in the nascent DNA, and the remaining bases are washed out. Each cycle a photo is taken and the base that has been incorporated by the polymerase is determined. Illumina and Qiagen use this technology. Both have preceded the sequencing process by an amplification step; spots of identical sequences are generated during this step, each spot can add a different nucleotide, which increases exponentially the throughput of the experiment [8, 9].

By other hand, the technology of addition of one single nucleotide does not use 3'-end blocked nucleotides. Instead, one nucleotide type is added in each cycle and the incorporation of such nucleotide to the DNA template is measured as light presence or as a change in pH. One important limitation of this technology is that several nucleotides of the same type can be incorporated during one cycle and the length of the homopolymers could be erroneously determined due to the limited detection power. Providers of this service are 454 and IonTorrent [10, 11].

2.3.3 Real-Time Sequencing of a Single Molecule

Most sequencing technologies generate short sequencing reads which length ranges from 35 bases to 700 bases. However, the new technologies which are classified as single molecule real-time sequencing can generate reads until 20,000 bases long with an error rate much more higher than any of the short reads technologies. PacBio and Oxford Nanopore Technology (ONT) use this technology. PacBio fixes the polymerase to the bottom of a well with transparent floor, through which an optical lector measure the fluorescence. Different values of fluorescence are recorded when different bases are incorporated into the nascent DNA. By other hand, ONT detects the incorporation of bases by the changes of voltage produced when different bases pass through a protein pore [12, 13].

2.3.4 Challenges

The generalized use of the different sequencing technologies in the last years has generated a massive amount of data. In consequence the size of genomic data is similar to the data produced by YouTube, Twitter, or astronomical science, which has been catalogued as big data domains [14]. This implies that in the near future, more efficient technologies must be developed to store, analyze, and distribute genomic data.

3 The Process of the Identification of Genomic Variants: Variant Calling

3.1 Introduction

The identification of the genomic positions at which an individual harbors a genetic variant is the first step to associate this genomic trait with a physical or metabolic characteristic, e.g., phenotypic trait. However, the correct identification of such positions is a challenge because the sequencing technologies produce short and imperfect reads which contain errors generated during the sequencing process.

The methods developed to identify genomic variants belong to two broad categories: the ones that need to know the genomic position from which every sequencing read was originated (alignment-based methods) and, by other hand, the ones that are based on changes in the frequency of a special type of words. In this chapter we will explain how the different methods work and how they use patterns to accomplish its goal.

3.2 The Importance of Genomic Variants

The variation of individuals at the genome level has been a research subject for a long time. The Project of the Human Genome found that the genome of each person contains between 250 and 300 mutations that are responsible for the loss of function of the corresponding protein and between 50 and 100 mutations that have been previously related with a hereditary disease [15]. Also, it has been discovered that each human population has an unique genetic fingerprint which implies that individuals with the same genetic ancestry share rare and common genetic variants [16].

Some genetic variants have been associated with physical traits such as eye color, height, or skin color; with biochemical capabilities such as ability to degrade some metabolite or particular compound; with syndromes and complex diseases such as cancer, obesity, diabetes, or cardiovascular disease; or even with complex traits such as intellectual ability.

3.3 Algorithms Based on Alignments

All the algorithms based on alignments that are used to pinpoint genomic variants rely on the same processing steps. First, the sequencing reads are aligned against a reference genome to identify the position from which each read was originated. Second, and optional, the experimental duplicates are removed. Third, the reads are realigned around the insertion or deletions that have been identified, and the quality scores per base are recalculated to improve the quality of the variants that are being identified. Finally, the genotype of each variant is assigned [17, 18].

3.3.1 Assignment of Origin Position Based on Pattern Recognition

This first step in the process of variant calling is one of the most exhaustive and time-consuming steps of the process. Two different types of algorithms are commonly used: the ones based on "hashes" and the ones based on "suffixes" along with the Burrows-Wheeler transform [19].

The premise of all hashes-based methods is to identify a shared substring between each sequencing read and the reference genome (seed) and to extend the alignment from this seed [20, 21]. Different implementations of this algorithm exist, and a lot of research has been focused on algorithm optimization aiming to reduce the required computing time or to improve the accuracy and sensibility. Some implementations have incorporated steps during the seeding phase. Some methods take advantage of global alignments strategies to complete the alignment around the seed [22, 23]. Some others have tried to use spaced seeds to anchor the alignment to allow sequencing errors between the seeds [24, 25]. Some implementations obtain all the overlapping substrings for each read and analyze only the reads that include a high percentage of substrings located on the genomic regions of interest [26]. Additionally, some effort has been devoted to reduce the extension phase by decreasing the number of genomic positions that must be interrogated; this has been achieved by comparing the base content of the read against all the possible origin locations and eliminating the genomic regions with highly dissimilar nucleotide content from the extension phase [27]. These methods are depicted in Fig. 1. The greatest disadvantage of this type of algorithms is that if the seed falls in a very repetitive region, so a big number of attempts will be done in the extension phase.

The suffixes of a word are all substrings that start in any position of such word and extend until the end of the word. All the suffixes of one specific word can be represented as paths from the root to one leaf in a suffix tree. By looking at this structure, it can be determined if a target substring exists in the original word. So, if there is a path root-leaf in the suffix tree of the original word that is read as the target substring, then the target substring exists in the original word. The main advantage of this structure is that it collapses all the repetitions of the same substring into a unique path root-leaf (see Fig. 2) [28]. By other side, the main disadvantage is that it only allows to find exact matches of the target substring. Also, it has been shown that suffix trees are extremely big structures. In the case of the human genome, the corresponding suffix tree cannot be stored in RAM memory and, as a consequence, it cannot be rapidly interrogated.

Finally, there is another data structure that collapses all sequence repetitions from a genome and that is highly compact which allows its storage and rapid interrogation, the Burrows-Wheeler transform (BWT). The generation of the BWT requires the computing of the corresponding suffix array. The procedure to create this structure is as follows: obtain all the suffixes from a word, sort them alphabet-

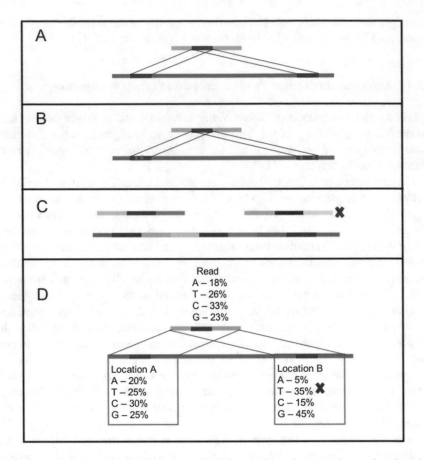

Fig. 1 Some algorithms used to assign a position of origin for each sequencing read are based on hashes. (**a**) During the seeding phase a shared substring (seed, red line) between the sequencing read (green line) and the reference genome (blue line) is identified. (**b**) The sensibility has been improved by using spaced seeds. (**c**) The computing time has been reduced during the seeding phase if only reads that share some fraction of substrings with the genomic region of interest are further analyzed (only some substrings for the read and reference genome sequences are shown as colored blocks) (**d**) or during the extension phase if only the genomic regions with base content similar to the read of interest are further analyzed, in this example only the genomic location A will be further analyzed

ically taking into account that the character that indicates the end of the word is the first one in alphabetic order, store the start position from each suffix, and ready: the suffix array of the original word has been created. As you can see, this array is no more than a list of positions sorted by the alphabetic order of all suffixes of the original word. This list goes from 1 to the length of the original word, which is around 3000 million in the case of the human genome. A circular array of suffixes must be generated to build the BWT: the whole word is written for each index of the suffix array, it must start from the position indicated in the array until the end

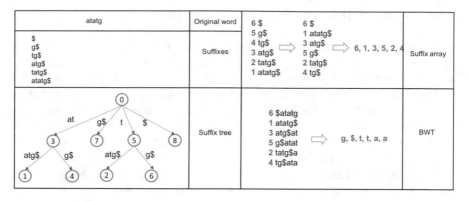

Fig. 2 Generation of the suffix tree, suffix array and Burrows-Wheeler Transform

of the original word, and it must continue by the beginning of the original word until one letter before the position indicated in the array. The BWT is formed by the last letter of the circular array of suffixes (see Fig. 2). Even though it is not easy to visualize, by using the BWT and the original word, the suffix array can be searched in the same way as it would a suffix tree, enabling to know if a target substring is present or not at the original word. This strategy has been widely adopted by diverse mapping algorithms [29].

3.3.2 Duplicate Removal

Any biological sample that will be sequenced must follow a preparation process. During this process the DNA is fragmented, the fragments are selected by size, and only those selected fragments are ligated to universal adapters by both ends. These adapters are used as primers to amplify the DNA by a PCR (polymerase chain reaction). This reaction produces thousands of copies of the original DNA. The PCR products are bound to the device in which the sequencing process will occur.

When the coverage of the sequencing experiment is low and the size of the sequencing fragments is large, it is highly improbable to sequence twice the same exact genomic fragment. So, if more than one sequencing read is aligned to the same position on the reference genome these reads are considered PCR duplicates and all but one are removed from the analysis [30, 31].

If this procedure is skipped, then a polymerase error or an allele found at this DNA fragment will be proportionally overrepresented in all subsequent steps of the variant calling process.

3.3.3 Realignment Around Insertions and Deletions

Mapping algorithms have been developed to find occurrences of highly similar substrings; these algorithms try to find the best possible alignment that fulfills some similarity criteria. Frequently, the presence of insertions and deletions in the sequenced samples can introduce errors during several steps of the variant calling process. Indeed, it has been observed that the alignments that contain insertions or deletions (indel) frequently include a high number of mismatches around the indel. These misaligned bases could be incorrectly identified as point mutations by posterior steps of the pipeline.

Most variant callers incorporate modules responsible to correct the alignments around the indels [18, 32]. These modules minimize the number of observed total changes by taking into account not only the read being analyzed but the whole set of sequenced reads. In this way, the read that contains the indel flawlessly mapped is used to correct the alignment of the rest of the reads. As a consequence, at the end of this step, most reads will support the existence of a consensus indel, e.g., an indel free of multiple substitutions at both sides.

The rightful identification of indels is a problem under research. Specialized algorithms have been developed with the only aim to accurately identify insertions and deletions. One of these methods assembles de novo all the reads mapped into a set of genomic regions using de Brujin graphs. These assemblies are then aligned against the reference genome by a gap-sensitive algorithm [33].

3.3.4 Quality Score Recalibration Using Sequence Patterns

Each sequencer assigns a quality to each of the bases it calls. This quality score is related to the probability that the base was incorrectly called. The quality values are assigned according to a Phred scale, which is defined as $Q = -10\log_{10} P$, where Q is the quality value of a certain base and P is the probability that the base has been called incorrectly; this means that a quality value of 10 implies that such base has a probability of 1 in 10 of being an error; a quality of 20 implies that base has a probability of 1 in 100 of being an error; and so on [34]. Also, the most common variant calling algorithm include a step to recalibrate these quality scores by base in order to improve the quality of the identified variants [18, 32].

During this recalibration step, the algorithms look for technical biases that the sequencer has not taken into account when assigning qualities. To achieve this, the algorithm needs a catalog of expected variation. In the case of humans, the 1000 genome project has generated a catalog with millions of common mutations in different populations. All the positions found at the catalog are labeled as expected variation. Subsequently, it is assumed that all those positions that vary and that have not been labeled as expected variation are an error introduced by the sequencer. Taking into account these two types of sites, a model is trained which takes into account the current quality of each base, the position within the read for each base, and the specific nature of the previous base and the current one. Using all of these variables as covariates, the model looks for patterns in which the sequencer commonly assigns higher or lower qualities than it should. In these regions, quality values are recalibrated.

By doing this, two types of errors are corrected: the quality of the bases is decreased where the sequences has overestimated the quality values, lowering the confidence of bases incorrectly classified as high confidence; and at the same time, the quality of the bases is increased where the sequencer has underestimated the quality, increasing the confidence of bases which would have been incorrectly ruled out [32].

3.3.5 Genotype Assignment

A genotype is assigned for those sites for which variation has been detected during this step of the variant calling process, e.g., the alleles that exist at each genomic position are determined. The genotype assignment process is not a direct process due to random fluctuations in the sequencing coverage for each one of the alleles present at each genomic position during the sequencing process; moreover the sequencing machines introduce errors resulting in some fraction of incorrect reads [35].

The first algorithm designed to assign a genotype was based on a very simple premise: all heterozygous sites should have a defined proportion of reads supporting the reference allele. All the sites with a higher or lower percentage of reads carrying the reference allele were classified as homozygous reference or homozygous no reference sites, respectively [36, 37]. Nevertheless, this approach has been proven to have two very important limitations: it tends to misassign genotype in low coverage regions, and it does not generate a quantitative measure of the assigned genotype.

More sophisticated algorithms solve this problem by treating the genotype assignment task as a problem of posterior probabilities calculation. In this approach, the sequencing data is used to calculate the posterior probability for each genotype at every genomic position given that specific sequencing data [38]. In addition to calculate the most probable genotype, this approach offers the advantage of calculating a quality value for the assigned genotype. Usually this quality value corresponds to the ratio of the probability of the most probable genotype between the probability of the second most common genotype. In this way, the quality value indicates how many times is more probable the assigned genotype than any other possible genotype [31].

There are several ways to increase the confidence of the assigned genotype. One way is to use observed allelic frequencies to obtain the anterior probability of such variant. This anterior probability can be used in the calculus of the posterior probability. In some cases, these prior probabilities could help to decide between equiprobable genotypes. The allelic frequencies seen in different human populations can be consulted in dbSNP [24, 38], or they could be obtained from the data generated in the same experiment [39].

3.4   Algorithms Not Based on Alignments

In the cases in which the process of variant calling should be done but there is not a high quality reference genome or when the comparison against it is considered counterproductive, the steps described in the previous section cannot be applied. For such applications, a great diversity of algorithms have been developed.

A first approximation is to assemble de novo the target genome. However, when the sequencing experiment is conducted in only one individual, this method results in an assembled genomic sequence instead of a list of variable positions. One of the most widely used algorithms for assembly genomes de novo is the Brujin graph. Let's remember that a graph is formed by a series of nodes joined by vertices. In the case of a genomic assembly Brujin graph, the nodes represent substrings, and the existence of a vertex joining two specific nodes indicates that the two substrings exist as contiguous words in the genome. In this way, a path through a Brujin graph represents a word that exists in the sequenced genome. In order to create a de Bruin graph from sequencing data, all substrings of size k-1 found at the reads are represented as a node, two nodes are connected only if there is a substring of size k in the reads composed by the prefix represented in the first node and the suffix represented in the second node. For example, let's assume there are two nodes that represent the ATGC and TGCT substrings; these two nodes will be connected if the ATGCT string exists in at least one read [40]. At the end, a graph that represents the complete genomic sequence of the sequenced organism is obtained, and the genomic variants can be represented as alternative paths in such graph.

Some other methods have been developed relying in the existence of a variation catalog. In one of them, a dictionary is created based on genomic substrings that contain the variants and some substrings located on the genomic neighborhood of the variants. The number of occurrences of the substrings that are part of the dictionary is counted. The regions with a coefficient of uniqueness close to 1 and that contain specific frequency patterns are localized. And finally, the neighborhood substrings are used as seeds to extend the alignment against the reference genome and identify the variation   [41]. In an alternative method, multiple substrings overlapping each variant are designed. Each variant is covered by k pairs of substrings (k is the size of the substrings), and each pair corresponds to the pair of alternative alleles for the variant position. The genotype for each site is obtained from the number of reads in which the substring that contains each allele exist [42].

Some algorithms have focused on reconstructing the variation for regions that are quite different from the reference. Once again, changes in the frequency of consecutive substrings in the reference genome are used as markers for the regions in which variation should be sought. For these regions, a reconstruction of the genome is made in the following way: the last substring from the reference found at a high frequency in the reads is localized, and each of the possible four nucleotides is added. Then, the nucleotide that produces the substring with the highest number of occurrences in the reads is assigned as the next base in the genomic assembly. If there are several bases that produce substrings with the same account, multiple possible versions of the genome are created from that point. This procedure is repeated until reaching the next substring identical to some region of the reference genome [43].

Finally, changes in the frequency of substrings located at only one position in the reference genome have been used to pinpoint variable regions of the genome. Also, the unique substrings bordering such regions have been used to attract the reads harboring the variant nucleotide [44].

4   Genetic Expression Profile for Diagnostic of Diseases

4.1   Background of Genetic Profile Construction

With the arrival of microarray technologies, massive sequencing, and omic sciences in general, specialists have much more information to perform global genetic expression analysis, but they are also dealing with the problem of summarizing and ordering the observed data to gather reliable and trustworthy information. Then, in this section, we analyze how to construct expression genetic profiles with the algorithm Gene Set Enrichment Analysis (GSEA), using microarray, high-throughput PCR or RNA sequencing (RNA-seq) data.

Furthermore, we discuss how genetic profiles expression can be linked with diseases or relevant phenotypes. Additionally, we explain how machine learning techniques, such as clustering methods, can be used to predict diagnostic or effective treatment for illness.

4.2   Methodologies for Genetic Profile Construction

We call genetic expression profiles to the set of genes that increase or decrease their expression in a consistent manner and that it has been linked to a phenotype, disease, or condition of the set of samples from which it was extracted. There are three prominent methodologies for constructing genetic profiles: high-throughput PCR, microarrays, and RNA-seq. These three methods have revolutionized the genetic expression paradigms and have enabled the generation of better gene expression profiles. High-throughput PCR has been applied for quantifying levels of specific genes. Using this technique, RNA is converted to cDNA and then amplified by the PCR. DNA microarrays make it possible to simultaneously access to the messenger RNA (mRNA) expression of thousands of genes.

Imaging processing from the microarray surface can generate a single vector that summarizes the expression of a group of genes under a condition, or the difference between samples with paired properties, such as the status of a disease or a treatment associated with any specific sample. When the data comes from a microarray, the quantification of the intensity of fluorescent signals in a cell indicates a change in the expression of the genes.

Finally, RNA-seq has become a very popular technique widely used for gene expression quantification. This technique, also called whole transcriptome shotgun sequencing (WTSS), uses next-generation sequencing (NGS) to reveal the presence and quantity of RNA in a biological sample at a given moment. When the data comes from an RNA-seq experiment, the quantification of the levels of expression is obtained counting the number of reads that aligns with each gene in the genome of interest. The list of genes can be obtained from previously annotated genomes or from the assembly de novo of unknown transcripts. The quantification process is made per sample, and each sample is labeled as belonging to a treatment, tissue, or condition of interest.

Several authors agree that it is difficult to find individual genes that fully explain a phenotype or disease. What happens more often is that a great number of genes make small contributions to that phenotype, which are not significant in a gene by gene context. Furthermore, when a genetic expression experiment is repeated, for example, when a new group of patients is sampled, the number of the genes with differential expression identified in the first experiment versus the genes with differential expression in the second one is usually very small. It is easier to detect subtle coordinated changes between the genes of each experiment than within the changes of just one genes. This is the reason for directing efforts toward the comparison of gene expression profiles more than the isolated gene analyses.

We speak about differential gene expression when the level of expression varies over a random difference threshold. Usually, the comparison between these levels of expression takes place between different conditions, tissues, or samples.

It is highly recommendable that each experiment has at least one replicate. In our experience, the experimental designs with at least three replicates per condition or sample allows to identify outliers, a process that would be impossible in the absence of replicates, resulting in the increase of the bias in the analysis. The practical utility of having replicas is that we can calculate the average of the values of expression and the variability among them. With this information, we may evaluate when the difference between two levels of expression is statistically reliable, in other words, that it is not produced randomly.

Most of the methods for differential expression analysis use a procedure similar to the following:

- Calculate the mean (or median) of expression levels for the gene in the group of replicates of each phenotype or tissue or condition.
- Calculate the difference between the means for each group
- Calculate a measure of the variability of the gene expression in the samples within each condition or tissue, such as the standard deviation of the expression within each group of replicates.
- Calculate an association score. For example, divide the difference between the two means by the measure of variability. If the difference between the two means is large, and the measure of variability is small, then the statistical association will

be large. A large association score gives evidence that the gene is differentially expressed between the two conditions or tissues.

GSEA was designed to work with the output of microarrays [45]. It is a statistical method to determine if predefined sets of genes are differentially expressed in different phenotypes. Predefined gene sets may come from any source, such as a known metabolic pathway, a Gene Ontology category, a set of co-expressed genes observed in a previous experiment, or any user-defined set. The results of this method have been reliable in many independent experiments [46, 47], and it is available in various websites [45, 48].

The GSEA method is described in [49]. Here, the steps of this method are briefly described in the following [45].

Step 1. Calculation of an Enrichment Score (ES): We calculate an association score for each gene, which measures the difference of the gene expression in the two phenotypes or conditions, using a t-test, Wilcoxon rank sum test, and signal-to-noise (SNR), among others.

After we calculate the association score for each of the N genes in the entire data set D, we sort the genes by their association score to produce a list L. If the gene set S is related to the phenotype, then we would expect that the genes in S will tend to have higher association scores than those in gene sets not related to the phenotype. The, the proportion of genes in S that rank near the top of the list L should be greater than the proportion of other genes. If we look at all the genes in L up to a position i, we can compare the fraction of the genes in S that are present before i with the fraction of all the N genes (except those in S) that are present also before i. If more genes from S occur upper in the list, then gene set S is enriched.In summary, we calculate the difference between (i) the fraction of the genes in S that are present before i and (ii) the fraction of all the N genes (except those in S) that are present before i across all possible positions in the list L. If there is no enrichment of the gene set S, then the ES should be near zero, that is, there is little difference in the proportion of genes in S that rank near the top of L and the proportion of other genes near the top of L.

Step 2. Estimation of the significance level of the ES: We estimate the statistical significance (nominal P value) of the ES by using an empirical phenotype-based permutation test procedure that preserves the complex correlation structure of the gene expression data.

Step 3. Adjustment for Multiple Hypothesis Testing: When an entire database of gene sets is evaluated, we adjust the estimated significance level to account for multiple hypothesis testing. For this, we calculate the false discovery rate (FDR).

As previously stated, with the advent of next-generation sequencing data, we can generate a table based on the quantification of the expression of each gene in each sample and its replicates. Then, it is possible to make differential expression analyses using methods implemented in R, such as DESeq2 [50], EdgeR [51], NOISEq [52], or Limma [53]. These methods use hypotheses tests to evaluate if the difference between two measurements is statistically significant or not, followed by adjusted multiple hypotheses testing.

There are tools in which it is very simple to make differential expression analyses based on the gene expression quantification, such as IDEAmex [54] (http://www.uusmb.unam.mx/ideamex/). This website allows the user to test simultaneously the DESeq, EdgeR, NOISEq, and Limma methods.

4.3 Considerations for Clustering Profiles

When the construction of genetic expression profiles has been achieved, the following step is to find similarity or dissimilarity measurements. Some widely used measurements include Pearson correlation coefficient, Euclidean distance, Spearman rank correlation, and chi-square-based measure per Poisson distributed data [55].Clustering methods have been widely used on gene expression data for categorizing genes with similar expression profiles, for example, hierarchical clustering [56], k-means [57], self-organizing maps [58], Gaussian and non-Gaussian clustering [59], or Mixture models [60]. These techniques have been used, for example, to diagnose cancer [61]. A good review of cDNA analysis methodologies to discover relationships between genotypes and phenotypes can be found in [62].

There are multiple efforts to make the classification of gene expression profiles a low-cost diagnostic tool. Although they are still being refined, there have been great advances. For example, it has been demonstrated that a classifier based on patient's blood RNA profile can distinguish between bacterial and viral infection [63].

4.4 Scope and Limitation

There is an implicit assumption that the expression values for all genes are essentially on the same scale. For this and other reasons, it is advisable to standardize the expression values for each gene before performing any GSEA analysis. However, in other cases, it is advisable to start the analysis from raw counts, because some algorithms, such as EdgeR, DESeq, or NOISEq, include a custom normalization stage.

In a RNA-seq experiment, genes with very low counts will be the origin of the majority of the incorrect or poorly reliable conclusions. Because of this, it is highly recommended to discard these genes from the analysis in early stages. It is also widely known that the predefined gene sets work poorly if the number of genes in the set is too small. Moreover, if the microarray experiment includes a small number of samples from each phenotype, then the permutation test for statistical significance will tend to give less accurate estimates of the p-values.

When we analyze gene expression data, we commonly compare two phenotypes (healthy versus diseased) or two tissues (liver vs. kidney) or two conditions (humidity and drought), but we know that other factors, such as age, gender, previous medical history, temperature, and so on, may also cause changes in gene expression. These additional factors usually increase the variability in gene expression within each sample. Then, if we do not control or adjust for these other factors, we may not be able to detect a significant effect due to the factor of interest (healthy, diseased, tissue, humity or drougth), or, in some cases, we could relate the observed effect to the incorrect factor.

Definitely, there are many open problems and not so fulfilling answers. However, it is a fact that each time there are more accurate, efficient, and inexpensive early diseases prognostics and identifications of specific treatments for each patient using gene expression profiles, which leads into the threshold of the personalized medicine.

5 Gene-Disease Association Extraction from Biomedical Literature

5.1 Background

Biomedical literature has increased rapidly during the past decades. In 2015 more than three thousand of scientific articles were published every day in different scientific journals [64]. As a result of this accelerated growth we have access to a huge amount of knowledge contained into this literature. If scientific community wants to build wider perspectives of biological phenomena for having a better understanding of them, the extraction and integration of this knowledge is crucial.

Manual curation of literature is still the common approach of knowledge extraction because it delivers high-quality pieces of knowledge with elevated precision. However, this is time-consuming, demanding, and eventually surpassed by the amount of published literature. Consequently, developing automatic approaches for knowledge extraction to assist manual curation becomes relevant. These approaches have to recognize diverse patterns in unstructured information, as biomedical literature comprises mainly text.

In the next sections, we broadly describe the area of biomedical natural language processing (BioNLP), where some pattern recognition approaches have been employed for extraction of associations between genes and diseases mentioned in scientific literature. Some of these approaches have been used also to assist curation of gene-disease association databases, such as DisGeNET and DISEASES.

5.1.1 Biomedical Natural Language Processing

Biomedical natural language processing, also known as Biomedical Text Mining, is an area dedicated to develop approaches for knowledge extraction from biomedical literature [64, 65]. To this aim, BioNLP joins methods of different areas of artificial intelligence, such as pattern recognition, machine learning, and natural language processing (NLP).

Overall, BioNLP includes four techniques for knowledge extraction: document classification, document clustering, information retrieval, and information extraction [66, 67]. In document classification, biomedical documents are classified in predefined categories, whereas in document clustering, documents are organized

automatically in groups with unknown categories. Information retrieval is dedicated
to methods for selecting the best documents from a document collection given an
input query. Information extraction is described in the following section as it is the
matter of the present chapter.

5.1.2 Information Extraction

Information extraction is devoted to create techniques for detecting and extracting
relevant entities and their associations from biomedical literature [68, 69]. Detecting
biomedical entities in documents, such as diseases, protein names, gene names,
drugs, or secondary effects, is performed by the *named-entity recognition* (NER)
task, which has been a long-term challenge in BioNLP. Once a name entity is
detected within a document, the next task is to assign to it a normalized (standard)
name and a unique identifier (ID), which is called *normalization* [70]. Recovering
interactions between only two entities is called *relation extraction* or *association
extraction*, and when there is a complex interaction, for example, with more than
two entities, this is named as *event extraction* [65].

The traditional approach for NER is dictionary-based recognition using lists of
entities (dictionaries) obtained from curated databases. The main disadvantage of
this approach is that names in dictionaries could not match with names mentioned
in documents since authors use abbreviations and short names. For example, the
gene *Transforming growth factor beta 1* could be mentioned in scientific articles
as *Transforming growth factor-beta1, TGF-b1, TGF-b(1),* or *TGFb-1.* Usually,
dictionary-based approaches are combined with pattern recognition and machine
learning approaches to increase performance [71].

BioNLP community has paid attention to several kinds of interactions, such
as gene-disease, drug-drug, chemical-protein, protein-protein, and protein-binding
site, among others. In molecular biology, some efforts have been made to extract
many of the regulatory interactions [72]. One example is the BioNLP Share Tasks
organized by the Association for Computational Linguistics (ACL) [73]. Numerous
approaches for relation and event extraction have been proposed, an overview of
them can be found in [65, 72, 74].

5.2   *Automatic Extraction of Gene-Disease Associations*

Curation of gene-disease associations from literature helps biomedical community
to centralize relevant knowledge in databases. Then, this knowledge is employed
by researchers to perform specific studies about genes or diseases. However, as
mentioned earlier, traditional curation is time-consuming and demanding, so several
approaches to extract automatically these associations have been proposed; many of
them are also used for other kinds of interactions.

5.2.1　Gene-Disease Association Databases

Here, we briefly describe some gene-disease association databases to emphasize the active interest in these resources. For example, one of the oldest databases is the Online Mendelian Inheritance in Man® (OMIM®, https://omim.org/), which includes information about genes associated with Mendelian disorders. The Human Gene Mutation Database (HGMD®, http://www.hgmd.cf.ac.uk/ac/index. php) reports associations between gene mutations and human inherited disease. The Comparative Toxicogenomics Database (ctd, http://ctdbase.org/) contains manually curated gene-disease associations, among other relations like chemical-gene and chemical-disease. Also, the Genetics Home Reference (GHR, https://ghr.nlm.nih. gov/) offers information about genes associated with health conditions.

Two databases that employ automatic extraction of gene-disease association from literature are DISEASES and DisGeNET. The former (https://diseases.jensenlab. org) integrates gene-disease associations extracted automatically from literature, manually curated associations, cancer mutation data, and genome-wide association studies (GWAS) obtained from other databases. The latter (http://www.disgenet. org/) gathers information from several resources about genes and variants associated with human diseases.

Finally, other interesting projects dedicated to predict latent interactions among genes and diseases in networks are the Heterogeneous Network Edge Prediction platform (HNEP, https://het.io/hnep/) and the Combining dATa Across species using Positive-Unlabeled Learning Techniques (CATAPULT) method [75].

5.2.2　Co-occurrence Pattern

One of the most simple approaches to detect associations between biological entities in text is a co-occurrence pattern, that is, if two entities tend to co-occur in the same textual context (sentence, paragraph, abstract), consequently they are related [65, 76]. This pattern has been employed to extract effectively associations between genes and diseases to feed the DISEASES database [77]. After a dictionary-based NER process, a score that takes into account sentence- and abstract-level co-occurrence of genes and diseases is calculated. This scoring scheme was previously employed for detecting protein-protein interactions [78].

This approach based on the co-occurrence of entities is suitable to any kind of task where identifying the direction of the association is not necessary, which is the case for gene-disease associations. An advantage is the lack of dependency on linguistic analysis or training data, so it could be implemented fast and flexibly. Moreover, this approach has been employed to discover indirect relations between biological entities [79].

However, a disadvantage of a co-occurrence pattern is the high number of false-positive associations that this could generate (low precision) due to the unrestricted combination of entities co-occurring; however, it could extract almost all expected associations (high recall). In addition, because this approach does not take into consideration semantic information, negative relations are not detected. In the following example, the co-occurrence approach would propose false associations between the genes OPRM1 and OPRD1 and the opioid dependence disease.

Example of positive and negative associations within the same sentence "Secondary analyses employing the narrower phenotype of opioid dependence (83 affected individuals) demonstrated association with SNPs in PENK and POMC, but not in OPRM1 or OPRD1." [PMID 17503481]

5.2.3   Rule-Based Approaches

Textual and linguistic patterns observed by curators in samples of sentences have been codified in manually elaborated rules and then employed to extract automatically associations between biomedical entities, such as mutation-disease [80] and gene-disease [81–84]. Although rule-based approaches are far from being automatic, we include this description for presenting a kind of patterns that have been learned with automatic approaches.

Selecting *trigger words* to be included in rules is a common practice in the rule-based approach. These words can be nouns or verbs employed by authors to express associations, such as the verb *associate*. Also, selection of key prepositions is usual. A simple set of rules can be seen in the following example reformulated from [80].

Example of rules manually elaborated to detect associations

1. $NP_1 VG\_passive\_voice\{ASSOCIATE\} with NP_2$
2. $NP_1 VG\_active\_voice\{ASSOCIATE\} with NP_2$
3. $NP\{ASSOCIATE\} of NP_1 with/and NP_2$
4. $NP\{ASSOCIATE\} between NP_1 and NP_2$

In the previous example, NP stands for *noun phrase* and VG for *verb group*, and subscripts indicate equality of NPs. The word inside brackets denotes a lexical element, so $\{ASSOCIATE\}$ stands for words like *associates, associated*, and association. Thus, the rule 1 allows to extract an association between *TGF-beta* and *idiopathic pulmonary fibrosis* worded as "TGF-beta has been associated with idiopathic pulmonary fibrosis," while rule 4 extracts the same association, but worded as "Association between TGF-beta and idiopathic pulmonary fibrosis."

Contrary to co-occurrence pattern approach, rule-based approach generates less false positives (high precision), but typically does not recover all expected associations (low recall) since it is not possible to formulate rules for all patterns.

5.2.4   Machine Learning Approaches

Several approaches based mainly on supervised learning have been proposed to extract associations between biomedical entities from literature, for example,

Conditional Random Fields [85], Support Vector Machines classifiers [86], Rule Induction [87], and recently Deep Neural Networks [74].

For gene-disease associations, the BeFree system [88] is widely known, because it is used to help curation of the DisGeNET database. It is based on the combination of the Shallow Linguistic Kernel (SLK) proposed in [89] with the Dependency Kernel (DEPK) proposed in [90]. The SLK employs shallow syntactic information, such as part-of-speech categories (noun, verb) and lemmas of the words at the left and right of a gene or disease. On the other hand, the DEPK utilizes those syntactic categories of the shortest path between the gene and the disease in the syntactic dependency tree of the sentence.

Another adopted supervised learning approach has been a decision tree binary classifier to extract disease-mutation associations using a combination of statistical, distance, and sentiment features [91]. Also, there are approaches including automatic classifiers to perform subtasks of the association extraction, such as a maximum entropy classifier to detect trigger words and their arguments [84].

5.2.5 Current Challenges

Possibly, the main current challenge in gene-disease association extraction is the scarcity of curated data [84, 92]. This fact affects not only machine learning approaches, as they do not have accurate curated training data to learn predictive patterns, but also other kind of approaches due to an absence of a reference for evaluation. In addition, this scarcity of curated data could explain the abundance of rule-based and co-occurrence approaches. However, the creation of large curated data is time-consuming and demanding.

Finally, it is important to emphasize that there is room for improvement not only for the extraction of gene-disease associations but also for the NER and normalization of genes and diseases in text. Thus, new pattern recognition and machine learning techniques are welcome to enhance the performance of these challenging tasks.

6 Conclusion

Nowadays, next-generation sequencing (NGS) technologies are the choice of preference to inspect the whole genomic content of an organism. The identification of genomic variants, a process known as variant calling, is one of the initial steps to understand the interplay between the genetic background of an organism and its associated physical traits.

As NGS generate a huge amount of data, pattern recognition techniques and statistical models are applied by several steps of the variant calling process. Shared words between the reference genome and the reads are located in order to assign a position of origin for the reads produced by the sequencer. The Burrows-

Wheeler transform has been commonly used for this purpose. The sensibility can be increased if differences are allowed between these common words, and the search space can be bounded if only regions with similar base content are interrogated. Also, in a subsequent step, the base quality recalibration model distinguishes sequencing patterns associated with a tendency from the sequencer to overestimate or subestimate the base quality metrics. By other side, some algorithms identify variants based on changes in the frequency of special words, e.g., words found only once in the genome or words associated with known variants. Furthermore, some of these algorithms find genomic variants by looking for multiple paths in a Brujin graph.

Another important step is the construction and analysis of differential expression genetic profiles. Techniques developed for microarrays in the analysis of RNA-seq have been retaken. The use of pattern recognition and clustering techniques have improved the creation of gene expression profiles which have been related with specific diseases.

The automatic extraction of gene-disease associations from biomedical literature conveys several challenging tasks for text analysis, such as named-entity recognition, entity normalization, and association extraction. Despite the effort of the BioNLP community, there is space for pattern recognition techniques to improve the results in association extraction, especially to tackle the low recall of rule-based approaches and the low precision of the approaches based on the co-occurrence pattern. By applying pattern recognition and machine learning techniques, we can improve also the amount of curated data to enlarge gene-disease association database. Eventually, this will let the genomics and life science community to develop new diagnosis and treatments.

References

1. International Human Genome Sequencing Consortium. (2004). Finishing the euchromatic sequence of the human genome. *Nature, 431*, 931–945.
2. Taub, F. E., DeLeo, J. M., & Thompson, B. E. (1983). Sequential comparative hybridizations analyzed by computerized image processing can identify and quantitate regulated RNAs. *DNA, 2*(4), 309–327.
3. Churchill, G. A. (2002). Fundamentals of experimental design for cDNA microarrays. *Nature Genetics, 32*, 490.
4. International Human Genome Sequencing Consortium. (2001). Initial sequencing and analysis of the human genome. *Nature, 409*, 860–921.
5. Shendure, J., Balasubramanian, S., Church, G. M., Gilbert, W., Rogers, J., Schloss, J. A., & Waterston, R. H. (2017). Dna sequencing at 40: Past, present and future. *Nature, 550*(7676), 345.
6. Zhang, J., Chiodini, R., Badr, A., & Zhang, G. (2011). The impact of next-generation sequencing on genomics. *Journal of Genetics and Genomics, 38*(3), 95–109.
7. Goodwin, S., McPherson, J. D., & McCombie, W. R. (2016). Coming of age: Ten years of next-generation sequencing technologies. *Nature Reviews Genetics, 17*(6), 333.
8. Drmanac, R., Sparks, A. B., Callow, M. J., Halpern, A. L., Burns, N. L., Kermani, B. G., Carnevali, P., Nazarenko, I., Nilsen, G. B., Yeung, G., et al. (2010). Human genome sequencing using unchained base reads on self-assembling dna nanoarrays. *Science, 327*(5961), 78–81.

9. Valouev, A., Ichikawa, J., Tonthat, T., Stuart, J., Ranade, S., Peckham, H., Zeng, K., Malek, J. A., Costa, G., McKernan, K., et al. (2008). A high-resolution, nucleosome position map of C. Elegans reveals a lack of universal sequence-dictated positioning. *Genome Research, 18*(7), 1051–1063.
10. Margulies, M., Egholm, M., Altman, W. E., Attiya, S., Bader, J. S., Bemben, L. A., Berka, J., Braverman, M. S., Chen, Y.-J., Chen, Z., et al. (2005). Genome sequencing in microfabricated high-density picolitre reactors. *Nature, 437*(7057), 376.
11. Rothberg, J. M., Hinz, W., Rearick, T. M., Schultz, J., Mileski, W., Davey, M., Leamon, J. H., Johnson, K., Milgrew, M. J., Edwards, M., et al. (2011). An integrated semiconductor device enabling non-optical genome sequencing. *Nature, 475*(7356), 348.
12. Clarke, J., Wu, H.-C., Jayasinghe, L., Patel, A., Reid, S., & Bayley, H. (2009). Continuous base identification for single-molecule nanopore dna sequencing. *Nature Nanotechnology, 4*(4), 265.
13. Eid, J., Fehr, A., Gray, J., Luong, K., Lyle, J., Otto, G., Peluso, P., Rank, D., Baybayan, P., Bettman, B., et al. (2009). Real-time dna sequencing from single polymerase molecules. *Science, 323*(5910), 133–138.
14. Stephens, Z. D., Lee, S. Y., Faghri, F., Campbell, R. H., Zhai, C., Efron, M. J., Iyer, R., Schatz, M. C., Sinha, S., & Robinson, G. E. (2015). Big data: Astronomical or genomical? *PLoS Biology, 13*(7), e1002195.
15. Consortium, G. P. (2010). A map of human genome variation from population-scale sequencing. *Nature, 467*(7319), 1061.
16. McVean, G., Altshuler (Co-Chair), D., Durbin (Co-Chair), R. et al. (2012). An integrated map of genetic variation from 1,092 human genomes. Nature 491, 56–65.
17. Nielsen, R., Paul, J. S., Albrechtsen, A., & Song, Y. S. (2011). Genotype and SNP calling from next-generation sequencing data. *Nature Reviews Genetics, 12*(6), 443.
18. DePristo, M. A., Banks, E., Poplin, R., Garimella, K. V., Maguire, J. R., Hartl, C., Philippakis, A. A., Del Angel, G., Rivas, M. A., Hanna, M., et al. (2011). A framework for variation discovery and genotyping using next-generation dna sequencing data. *Nature Genetics, 43*(5), 491.
19. Schbath, S., Martin, V., Zytnicki, M., Fayolle, J., Loux, V., & Gibrat, J.-F. (2012). Mapping reads on a genomic sequence: An algorithmic overview and a practical comparative analysis. *Journal of Computational Biology, 19*(6), 796–813.
20. Langmead, B., Trapnell, C., Pop, M., & Salzberg, S. L. (2009). Ultrafast and memory-efficient alignment of short dna sequences to the human genome. *Genome Biology, 10*(3), R25.
21. Langmead, B., & Salzberg, S. L. (2012). Fast gapped-read alignment with Bowtie 2. *Nature Methods, 9*(4), 357.
22. Lunter, G., & Goodson, M. (2011). Stampy: A statistical algorithm for sensitive and fast mapping of Illumina sequence reads. *Genome Research, 21*(6), 936–939.
23. Ning, Z., Cox, A. J., & Mullikin, J. C. (2001). SSAHA: A fast search method for large dna databases. *Genome Research, 11*(10), 1725–1729.
24. Li, R., Yu, C., Li, Y., Lam, T.-W., Yiu, S.-M., Kristiansen, K., & Wang, J. (2009). Soap2: An improved ultrafast tool for short read alignment. *Bioinformatics, 25*(15), 1966–1967.
25. Jiang, H., & Wong, W. H. (2008). SeqMap: Mapping massive amount of oligonucleotides to the genome. *Bioinformatics, 24*(20), 2395–2396.
26. David, M., Dzamba, M., Lister, D., Ilie, L., & Brudno, M. (2011). Shrimp2: Sensitive yet practical short read mapping. *Bioinformatics, 27*(7), 1011–1012.
27. Rizk, G., & Lavenier, D. (2010). Gassst: Global alignment short sequence search tool. *Bioinformatics, 26*(20), 2534–2540.
28. Rivals, E., Salmela, L., Kiiskinen, P., Kalsi, P., & Tarhio, J. (2009). *Mpscan: Fast localisation of multiple reads in genomes* (In: Salzberg S.L., Warnow T. (eds) algorithms in bioinformatics. WABI 2009. Pp. 246–260. Lecture notes in computer science, vol 5724). Berlin, Heidelberg: Springer.
29. Li, H., & Durbin, R. (2010). Fast and accurate long-read alignment with burrows–wheeler transform. *Bioinformatics, 26*(5), 589–595.
30. Li, H., Handsaker, B., Wysoker, A., Fennell, T., Ruan, J., Homer, N., Marth, G., Abecasis, G., & Durbin, R. (2009). The sequence alignment/map format and samtools. *Bioinformatics, 25*(16), 2078–2079.

31. Li, H. (2011). A statistical framework for snp calling, mutation discovery, association mapping and population genetical parameter estimation from sequencing data. *Bioinformatics, 27*(21), 2987–2993.

32. McKenna, A., Hanna, M., Banks, E., Sivachenko, A., Cibulskis, K., Kernytsky, A., Garimella, K., Altshuler, D., Gabriel, S., Daly, M., et al. (2010). The genome analysis toolkit: A mapreduce framework for analyzing next-generation dna sequencing data. *Genome Research, 20*(9), 1297–1303.

33. Narzisi, G., O'rawe, J. A., Iossifov, I., Fang, H., Lee, Y.-h., Wang, Z., Wu, Y., Lyon, G. J., Wigler, M., & Schatz, M. C. (2014). Accurate de novo and transmitted indel detection in exome-capture data using microassembly. *Nature Methods, 11*(10), 1033.

34. Ewing, B., Hillier, L., Wendl, M. C., & Green, P. (1998). Base-calling of automated sequencer traces usingphred. i. accuracy assessment. *Genome Research, 8*(3), 175–185.

35. Ross, M. G., Russ, C., Costello, M., Hollinger, A., Lennon, N. J., Hegarty, R., Nusbaum, C., & Jaffe, D. B. (2013). Characterizing and measuring bias in sequence data. *Genome Biology, 14*(5), R51.

36. Harismendy, O., Ng, P. C., Strausberg, R. L., Wang, X., Stockwell, T. B., Beeson, K. Y., Schork, N. J., Murray, S. S., Topol, E. J., Levy, S., et al. (2009). Evaluation of next generation sequencing platforms for population targeted sequencing studies. *Genome Biology, 10*(3), R32.

37. Wang, J., Wang, W., Li, R., Li, Y., Tian, G., Goodman, L., Fan, W., Zhang, J., Li, J., Zhang, J., et al. (2008). The diploid genome sequence of an asian individual. *Nature, 456*(7218), 60.

38. Li, R., Li, Y., Fang, X., Yang, H., Wang, J., Kristiansen, K., & Wang, J. (2009). Snp detection for massively parallel whole-genome resequencing. *Genome Research, 19*(6), 1124–1132.

39. Martin, E. R., Kinnamon, D., Schmidt, M. A., Powell, E., Zuchner, S., & Morris, R. (2010). Seqem: An adaptive genotype-calling approach for next-generation sequencing studies. *Bioinformatics, 26*(22), 2803–2810.

40. Compeau, P. E., Pevzner, P. A., & Tesler, G. (2011). How to apply de Bruijn graphs to genome assembly. *Nature Biotechnology, 29*(11), 987.

41. Kimura, K., & Koike, A. (2015). Ultrafast snp analysis using the burrows–wheeler transform of short-read data. *Bioinformatics, 31*(10), 1577–1583.

42. Pajuste, F.-D., Kaplinski, L., Möls, M., Puurand, T., Lepamets, M., & Remm, M. (2017). Fastgt: An alignment-free method for calling common SNVs directly from raw sequencing reads. *Scientific Reports, 7*(1), 2537.

43. Audano, P., Ravishankar, S., & Vannberg, F. (2017). Mapping-free variant calling using haplotype reconstruction from k-mer frequencies. *Bioinformatics, 10*, 1659–1665.

44. Gómez-Romero, L., Palacios-Flores, K., Reyes, J., García, D., Boege, M., Dávila, G., Flores, M., Schatz, M. C., & Palacios, R. (2018). Precise detection of de novo single nucleotide variants in human genomes. *Proceedings of the National Academy of Sciences, 115*(21), 5516–5521.

45. Subramanian, A., Tamayo, P., Mootha, V. K., Mukherjee, S., Ebert, B. L., Gillette, M. A., Paulovich, A., Pomeroy, S. L., Golub, T. R., Lander, E. S., et al. (2005). Gene set enrichment analysis: A knowledge-based approach for interpreting genome-wide expression profiles. *Proceedings of the National Academy of Sciences, 102*(43), 15545–15550.

46. Patti, M. E., Butte, A. J., Crunkhorn, S., Cusi, K., Berria, R., Kashyap, S., Miyazaki, Y., Kohane, I., Costello, M., Saccone, R., et al. (2003). Coordinated reduction of genes of oxidative metabolism in humans with insulin resistance and diabetes: Potential role of PGC1 and NRF1. *Proceedings of the National Academy of Sciences, 100*(14), 8466–8471.

47. Petersen, K. F., Dufour, S., Befroy, D., Garcia, R., & Shulman, G. I. (2004). Impaired mitochondrial activity in the insulin-resistant offspring of patients with type 2 diabetes. *New England Journal of Medicine, 350*(7), 664–671.

48. Rubin, E. (June 2006). Circumventing the cut-off for enrichment analysis. *Briefings in Bioinformatics, 7*(2), 202–203. https://doi.org/10.1093/bib/bbl013.

49. Shi, J., & Walker, M. G. (2007). Gene set enrichment analysis (GSEA) for interpreting gene expression profiles. *Current Bioinformatics, 2*(2), 133–137.

50. Love, M. I., Huber, W., & Anders, S. (2014). Moderated estimation of fold change and dispersion for RNA-seq data with DESeq2. *Genome Biology, 15*(12), 550.
51. Robinson, M. D., McCarthy, D. J., & Smyth, G. K. (2010). edger: A bioconductor package for differential expression analysis of digital gene expression data. *Bioinformatics, 26*(1), 139–140.
52. Tarazona, S., García-Alcalde, F., Dopazo, J., Ferrer, A., & Conesa, A. (2011). Differential expression in rna-seq: A matter of depth. *Genome Research, 21*(12), 2213–2223.
53. Ritchie, M. E., Phipson, B., Wu, D., Hu, Y., Law, C. W., Shi, W., & Smyth, G. K. (2015). limma powers differential expression analyses for rna-sequencing and microarray studies. *Nucleic Acids Research, 43*(7), e47–e47.
54. Jiménez-Jacinto, V., Sánchez-Flores, A., & Vega-Alvarado, L. (2019). Integrative differential expression analysis for multiple experiments (ideamex): A web server tool for integrated rna-seq data analysis. *Frontiers in Genetics, 10*, 279.
55. Cai, L., Huang, H., Blackshaw, S., Liu, J. S., Cepko, C., & Wong, W. H. (2004). Clustering analysis of sage data using a poisson approach. *Genome Biology, 5*(7), R51.
56. Johnson, S. C. (1967). Hierarchical clustering schemes. *Psychometrika, 32*(3), 241–254.
57. Hartigan, J. A. (1975). *Clustering algorithms.* New York: John Wiley & Sons, Inc.
58. Tamayo, P., Slonim, D., Mesirov, J., Zhu, Q., Kitareewan, S., Dmitrovsky, E., Lander, E. S., & Golub, T. R. (1999). Interpreting patterns of gene expression with self-organizing maps: Methods and application to hematopoietic differentiation. *Proceedings of the National Academy of Sciences, 96*(6), 2907–2912.
59. Banfield, J. D., & Raftery, A. E. (1993). Model-based gaussian and non-gaussian clustering. *Biometrics, 49*, 803–821.
60. McLachlan, Geoffrey J., Basford, & Kaye E. (1988). "Mixture models : inference and applications to clustering", New York, United States: Marcel Dekker. Vol 84. P. 253 *p : ill.*
61. Lin, X., Afsari, B., Marchionni, L., Cope, L., Parmigiani, G., Naiman, D., & Geman, D. (2009). The ordering of expression among a few genes can provide simple cancer biomarkers and signal brca1 mutations. *BMC Bioinformatics, 10*(1), 256.
62. Hwang, B., Lee, J. H., & Bang, D. (2018). Single-cell rna sequencing technologies and bioinformatics pipelines. *Experimental & Molecular Medicine, 50*(8), 96.
63. Lopez, R., Wang, R., & Seelig, G. (2018). A molecular multi-gene classifier for disease diagnostics. *Nature Chemistry, 10*(7), 746–754.
64. Huang, C.-C., & Lu, Z. (2016). Community challenges in biomedical text mining over 10 years: Success, failure and the future, *Briefings in Bioinformatics, 17*(1), 132–44.
65. Ananiadou, S., Thompson, P., Nawaz, R., McNaught, J., & Kell, D. B. (2014). Event-based text mining for biology and functional genomics. *Briefings in Functional Genomics, 14*(3), 213–230.
66. Weiss, S. M., Indurkhya, N., Zhang, T., & Damerau, F. (2005). *Text mining: Predictive methods for analyzing unstructured information.* Publisher Springer-Verlag New York.
67. Manning, C. D., Raghavan, P., & Schütze, H. (2008). *Introduction to information retrieval.* Cambridge: Cambridge University Press.
68. Ananiadou, S., & McNaught, J., *Text mining for biology and biomedicine.* London: Artech House (2006).
69. Liu, F., Chen, J., Jagannatha, A. N., & Yu, H. (2016). Learning for biomedical information extraction: Methodological review of recent advances. *CoRR*, abs/1606.07993.
70. Cho, H., Choi, W., & Lee, H. (2017). A method for named entity normalization in biomedical articles: Application to diseases and plants. *BMC Bioinformatics, 18*(1), 451.
71. Basaldella, M., Furrer, L., Tasso, C., & Rinaldi, F. (2017). Entity recognition in the biomedical domain using a hybrid approach. *Journal of Biomedical Semantics, 8*(1), 51.
72. Vanegas, J. A., Matos, S., González, F., & Oliveira, J. L. (2015). An overview of biomolecular event extraction from scientific documents. *Computational and mathematical methods in medicine, Volume, 2015.*, Article ID 571381, 1–19. https://doi.org/10.1155/2015/571381

73. Chaix, E., Dubreucq, B., Fatihi, A., Valsamou, D., Bossy, R., Ba, M., Deléger, L., Zweigen-baum, P., Bessieres, P., Lepiniec, L., et al. (2016). Overview of the regulatory network of plant seed development (SeeDev) task at the BioNLP shared task 2016. In *Proceedings of the 4th BioNLP Shared Task Workshop* (pp. 1–11). ACL.
74. Björne, J., & Salakoski, T. (2018). Biomedical event extraction using convolutional neural networks and dependency parsing. In *Proceedings of the BioNLP 2018 workshop* (pp. 98–108).
75. Singh-Blom, U. M., Natarajan, N., Tewari, A., Woods, J. O., Dhillon, I. S., & Marcotte, E. M. (2013). Prediction and validation of gene-disease associations using methods inspired by social network analyses. *PloS One, 8*(5), e58977.
76. Chang, J. T., & Altman, R. B. (2004). Extracting and characterizing gene–drug relationships from the literature. *Pharmacogenetics and Genomics, 14*(9), 577–586.
77. Pletscher-Frankild, S., Pallejà, A., Tsafou, K., Binder, J. X., & Jensen, L. J. (2015). Diseases: Text mining and data integration of disease-gene associations. *Methods, 74*, 83–89.
78. Franceschini, A., Szklarczyk, D., Frankild, S., Kuhn, M., Simonovic, M., Roth, A., Lin, J., Minguez, P., Bork, P., Von Mering, C., et al. (2012). String v9. 1: Protein-protein interaction networks, with increased coverage and integration. *Nucleic Acids Research, 41*(D1), D808–D815.
79. Tsuruoka, Y., Miwa, M., Hamamoto, K., Tsujii, J., & Ananiadou, S. (2011). Discovering and visualizing indirect associations between biomedical concepts. *Bioinformatics, 27*, i111–i119, 06.
80. Mahmood, A. A., Wu, T.-J., Mazumder, R., & Vijay-Shanker, K. (2016). Dimex: A text mining system for mutation-disease association extraction. *PloS One, 11*(4), e0152725.
81. Rindflesch, T. C., Libbus, B., Hristovski, D., Aronson, A. R., & Kilicoglu, H. (2003). Semantic relations asserting the etiology of genetic diseases. In *AMIA Annual Symposium Proceedings* (Vol. 2003, p. 554). American Medical Informatics Association.
82. Masseroli, M., Kilicoglu, H., Lang, F.-M., & Rindflesch, T. C. (2006). Argument-predicate distance as a filter for enhancing precision in extracting predications on the genetic etiology of disease. *BMC Bioinformatics, 7*(1), 291.
83. Greco, I., Day, N., Riddoch-Contreras, J., Reed, J., Soininen, H., Kłoszewska, I., Tsolaki, M., Vellas, B., Spenger, C., Mecocci, P., Wahlund, L.-O., Simmons, A., Barnes, J., & Lovestone, S. (2012). Alzheimer's disease biomarker discovery using in silico literature mining and clinical validation. *Journal of Translational Medicine, 10*, 217.
84. Verspoor, K. M., Heo, G. E., Kang, K. Y., & Song, M. (2016). Establishing a baseline for literature mining human genetic variants and their relationships to disease cohorts. *BMC Medical Informatics and Decision Making, 16*(1), 68.
85. Bundschus, M., Dejori, M., Stetter, M., Tresp, V., & Kriegel, H.-P. (2008). Extraction of semantic biomedical relations from text using conditional random fields. *BMC Bioinformatics, 9*(1), 207.
86. Björne, J., Heimonen, J., Ginter, F., Airola, A., Pahikkala, T., & Salakoski, T. (2009). Extracting complex biological events with rich graph-based feature sets. In *Proceedings of the Workshop on Current Trends in Biomedical Natural Language Processing: Shared Task* (pp. 10–18). Association for Computational Linguistics.
87. Liu, H., Hunter, L., Kešelj, V., & Verspoor, K. (2013). Approximate subgraph matching-based literature mining for biomedical events and relations. *PLoS One, 8*(4), e60954.
88. Bravo, À., Piñero, J., Queralt-Rosinach, N., Rautschka, M., & Furlong, L. I. (2015). Extraction of relations between genes and diseases from text and large-scale data analysis: Implications for translational research. *BMC Bioinformatics, 16*(1), 55.
89. Giuliano, C., Lavelli, A., & Romano, L. (2006). Exploiting shallow linguistic information for relation extraction from biomedical literature. In *11th Conference of the European Chapter of the Association for Computational Linguistics*.
90. Kim, S., Yoon, J., & Yang, J. (2007). Kernel approaches for genic interaction extraction. *Bioinformatics, 24*(1), 118–126.

91. Singhal, A., Simmons, M., & Lu, Z. (2016). Text mining for precision medicine: Automating disease-mutation relationship extraction from biomedical literature. *Journal of the American Medical Informatics Association, 23*(4), 766–772.
92. Thompson, P., & Ananiadou, S. (2017). Extracting gene-disease relations from text to support biomarker discovery. In *Proceedings of the 2017 International Conference on Digital Health* (pp. 180–189). ACM.

Images Analysis Method for the Detection of Chagas Parasite in Blood Image

Leticia Vega-Alvarado, Alberto Caballero-Ruiz, Leopoldo Ruiz-Huerta, Francisco Heredia-López, and Hugo Ruiz-Piña

Abstract Chagas disease is caused by the protozoan parasite *Trypanosoma cruzi* (*T. cruzi*) and represents a major public health problem in Latin America. The most widely used technique for determining the development stage of Chagas disease is visual microscopical evaluation of stained blood smears. However, this is a tedious and time-consuming task that requires a trained operator. In this work, a system for the automatic parasite detection in stained blood smears images is proposed. The system includes a microscope with a specific automated positioning stage, for holding and moving slides under the microscope; a computer for controlling the stage position; and a digital camera to acquire images through the microscope. Such an image was analyzed in order to detect the parasite by means of image processing techniques. Experimental results show that it is feasible to have an automated system for the detection of the *Trypanosoma cruzi* parasite.

Keywords Automatic detection · Chagas diseases · Image processing · Microscope

L. Vega-Alvarado (✉)
Instituto de Ciencias Aplicadas y Tecnología, Universidad Nacional Autónoma de México (ICAT, UNAM), Ciudad de México, México
e-mail: leticia.vega@icat.unam.mx; alberto.caballero@icat.unam.mx; leoruiz@unam.mx

A. Caballero-Ruiz · L. Ruiz-Huerta
Instituto de Ciencias Aplicadas y Tecnología, Universidad Nacional Autónoma de México (ICAT, UNAM), Ciudad de México, México

National Laboratory for Additive and Digital Manufacturing (MADiT), Coyoacán, Cd. México, México
e-mail: hlopez@correo.uady.mx; rpina@correo.uady.mx

F. Heredia-López · H. Ruiz-Piña
Centro de Investigaciones Regionales Dr. Hideyo Noguchi, Universidad Autónoma de Yucatán, Mérida, Yucatán, CP, México

© Springer Nature Switzerland AG 2020
M. R. Ortiz-Posadas (ed.), *Pattern Recognition Techniques Applied to Biomedical Problems*, STEAM-H: Science, Technology, Engineering, Agriculture, Mathematics & Health, https://doi.org/10.1007/978-3-030-38021-2_3

1   Introduction

Chagas disease (also known as American trypanosomiasis) is a potentially life-threatening disease caused by the protozoan parasite *T. cruzi* (Fig. 1). According to the World Health Organization [1], it is found mainly in Latin American countries, where it is mostly transmitted to humans by a triatomine bug, also known as a "kissing bug." *T. cruzi* can be transmitted to a wide variety of wild or domestic mammals acting as host and reservoir. Infection occurs when the feces of triatomines contaminated with T. cruzi come into contact with mucous membranes. Often, the bugs defecate on the host while feeding, and the infected fecal droplets may be inadvertently passed to the skin lesions, mucosas of the eye, nose, or mouth [2, 3]. Transmission through blood transfusion can also easily occur and is the second major mean of infection [4]. An estimated eight million people are infected worldwide, mostly in Latin America [6]. It is estimated that over 10,000 people die every year from clinical manifestations of Chagas disease and more than 25 million people risk acquiring the disease [1].

The Chagas disease presents three main stages in humans [4, 5]. The acute stage, which is usually asymptomatic and unrecognized, is characterized for a high number of circulating parasites in the blood. In this stage 5% of infected patients may show fever, headache, drowsiness, tachycardia, edema, and shortness of breath [7]. When the Chagas disease is diagnosed early in this phase and a treatment is initiated, the patient can be cured [8]. Acute infection is succeeded by an asymptomatic intermediate stage that may last 5–40 years, during which the immune system appears able to reduce circulating trypanosomes to below microscopically detectable levels. The chronic phase of the disease develops in about 30–40% of infected patients. The primary targets are the heart and to a lesser extent the gut and nervous tissues [4].

Some tests can be useful for making a diagnosis, depending on the phase of the disease. The parasite detection by microscopy in fresh blood specimen, stained

Fig. 1 Trypanosoma cruzi
stained blood smears. CDC
Safer-Healthier-People
(http://www.dpd.cdc.gov)

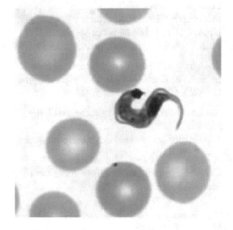

blood smears, buffy coat preparation, or serumen precipitate is of great importance in the acute phase [9]. However, this method is very tedious and time-consuming, and the accuracy of it depends on the operator's expertise. In addition, the time invested in the search for the parasite in blood smears increases exponentially in experimental studies whose purpose is to know the evolution of the infection. To address this problem, this work shows the design and implementation of an automated system for acquiring, analyzing, and classifying images coming from stained blood smears applied to the automatic *T. cruzi* detection.

2 Materials and Methods

The proposed system comprises two main stages: (1) stained blood smears positioning and image acquisition and (2) image analysis and classifying. The developed system was mounted on a light microscope and includes a system which automatically controls the movement of microscope stage in three directional axes [10]. The images are acquired with a digital camera that is installed at the top of microscope. The captured images are analyzed by digital image processing techniques to detect the *T. cruzi* parasites.

2.1 Automated System for Positioning and Image Acquisition

The automated motion system has a gear-rack system actuated by a three stepping motors (one per axis), which moves the platen on X and Y direction and adjust the focus (Z axis) every visual field. The number of stages (visual fields) are defined by the magnification zoom of the images, in this case magnification of 40× and resolution of 2500 × 1900 pixels. In order to decreases the mechanical backlash of the system, helycoidal gears were used. Every motor has a resolution of 200 steps per revolution, and the mechanical resolution of the system is 172 μm per motor step.

The image acquisition is performed with a 5 megapixel digital color CCD camera (Evolution MP, Media Cybernetics Inc.). The camera is coupled to the microscope by means of a C-mount optical adapter that is used to physically connect the digital camera at the top of the microscope. To obtain better illumination and contrast in the images, the conventional microscope light was replaced by LED white light. Figure 2 shows the camera coupled to the microscope. The images are acquired in RGB format with a resolution of 2500 × 1900 pixels and using a 40× objective. The acquisition and storage of the images is done automatically using a program developed in LABView©.

Fig. 2 Microscope-mounted
image acquisition system

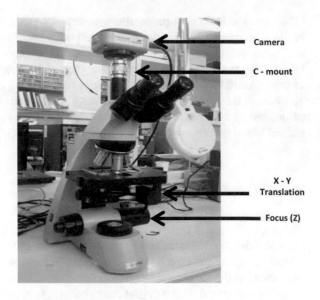

2.2 Images Analysis and Processing

The proposed methodology to detect the parasite in blood smears involves three stages. The first stage is used to improve the acquired images (image pre-processing) and eliminate noise. The second stage aims to detect the objects that are in the image (image segmentation), using the CIE-lab [11] color model and morphological characteristics. Finally, the third stage consists in the detection of the parasite [12].

2.3 Image Pre-processing

In order to eliminate noise, the acquired images were pre-processed, and the objects of interest were highlighted, in this case the *T. cruzi* parasite. The image pre-processing consisted of the automatic sequential application of filters and operators to the original image. This was achieved using the OpenCV image processing libraries [13]. A median filter with a 3×3 square kernel [14] was applied to remove the "salt and pepper" noise [15]. The kernel used is shown in equation 1. Later, a morphological "dilatation" operation to obtain more separate edges was performed.

$$K = \begin{bmatrix} & 1 & \\ 1 & \cdot 1 & 1 \\ & 1 & \end{bmatrix} \tag{1}$$

Fig. 3 Image pre-processing
steps

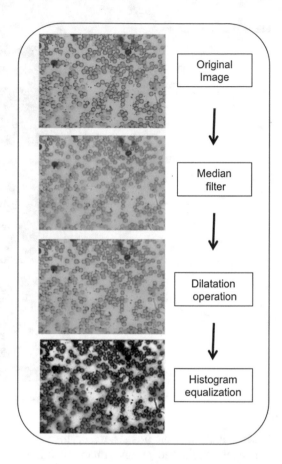

Finally, the image contrast is enhanced using histogram equalization, in order to contrast the color of the parasite. The steps of the image pre-processing are shown in Fig. 3.

2.4 Segmentation

Once the acquired images were pre-processed, it was essential to detect the parasite from the background. The techniques used to identify the objects of interest are usually referred to as segmentation techniques, as they extract the foreground from the background [16]. In this sense, a threshold algorithm to eliminate the background of the image was applied. All pixels with a hue value less than or equal to α threshold change their value to zero, that is, they become black. To set the α threshold value, a set of 30 images was randomly selected, and the average background color (α threshold) of the images was manually obtained. Subsequently,

Fig. 4 Segmentation steps

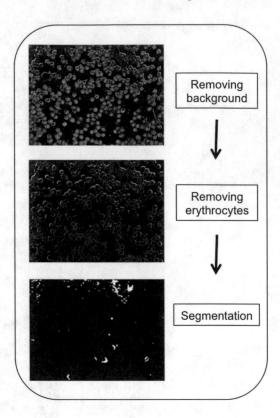

because in most of the images, the erythrocytes are abundant, a threshold algorithm was again applied to remove them. In this case, a pixel value was transformed to zero if the average of the pixel values in its neighborhood (3×3) was less than β threshold, where β was the average color value of the erythrocytes. Then a segmentation process using the CIE-lab (L for the lightness and a and b for the green–red and blue–yellow color components) color space was applied to obtain the regions of interest. This space color was used considering its advantage over the RGB model, such as good compatibility with human color perception and the possibility of using only one color feature, a or b, for segmentation purposes. To obtain the regions of interest, the pixels of the image were grouped considering certain ranges of tones, in our case we only grouped the pixels that had a hue similar to that of the parasite. Figure 4 shows the steps of the segmentation process.

2.5 Parasite Detection

Once the segmentation process was completed, features were extracted to detect the parasite. The method followed to select appropriate object features was to

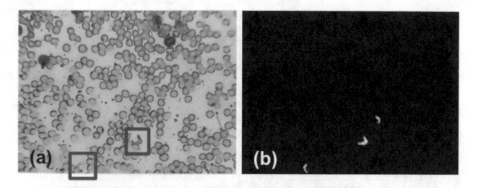

Fig. 5 Parasite detection. (**a**) Original image, (**b**) final parasite detection

calculate features involved during image annotation by a human expert. In this sense, the human expert identifies the parasite mainly by the shape, color, and size. In consequence, the following features were considered for the parasite detection: roundness (R) [17], hue mean (HM), and area (A). The values of these characteristics were obtained for each segmented object, by means of image processing tools in openCV.

In a first stage, large segmented objects were eliminated using an outlier exclusion criterion as follows. All objects with an area (A) outside the range [MPA $- 2\sigma$, MPA $+ 2\sigma$] were discarded, where MPA represents the average area of the parasite. Subsequently, objects out of hue range [AHP $- 2\sigma$, AHP $+ 2\sigma$] were deleted, where AHP represents the average hue of the parasite. Finally, considering the roundness feature (R), objects outside the range [MR $- \sigma$, MR $+ \sigma$] were eliminated, where MR represents the average roundness of the parasite. In all ranges σ represents the standard deviation. Hence, every object in the image will be classified as parasite if their three feature values are within each range. Figure 5 shows the final parasite detection.

3 Results

The mechanical characterization of the positioning system was made by means of a laser interferometer. The linear resolution of positioning system was 172.8 μm per motor step, with a backlash of 100 μm. Such values were implemented in the control system in order to increase the precision of the system. One of the major disadvantages of the proposed device is the necessity to implement an autofocus routine every ten images acquired. This represents too much time between every visual field acquired.

On the other hand, 500 images of a blood smear were automatically acquired to evaluate the performance of the image processing and analysis methodology. Of

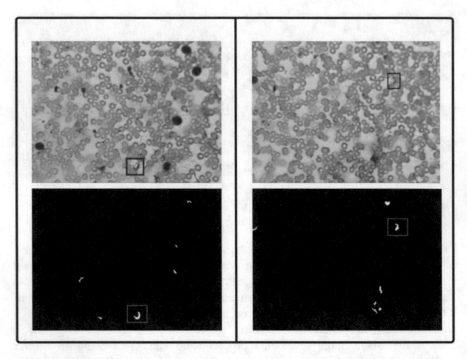

Fig. 6 Parasite detection complexity. (**a**) Medium complexity, (**b**) High complexity

these images, only the first ones taken after the focus process (every ten images acquired) were selected, in order to have the most focused ones and the best. Hence, only 50 images were analyzed. We observed that there were three levels of complexity in the segmentation process to detect the parasite. These complexity levels depend on the amount of erythrocytes in the sample. The images of (1) low complexity were those where the parasites were not in contact with any erythrocyte, (2) medium complexity were those where the parasite was in contact with erythrocytes, and (3) high complexity were those where the parasites were found within an agglomerate of erythrocytes. In cases (1) and (2), the parasites could be segmented by means of the proposed methodology. However, for complexity (3), the proposed methodology was not able to segment the parasite without the erythrocytes. Figure 6 shows two image with medium and high levels of complexity.

Moreover, the performance of the parasite detection process, was evaluated for the three levels of complexity. The results are summarized in Table 1. All correctly recognized parasites represent the true positive (TP), all the objects misclassified as parasite represent the false positive (FP), and all the parasite not classified as parasite represent the false negative (FN). The total number of parasites (TotPar) in the three complexity levels was manually quantified by the expert in the 50 test images.

Table 1 Parasite detection performance

Complex level	TotPar	TP	FP	FN
Low	23	23		0
Medium	11	8	3	3
High	9	0		9

TotPar Total number of parasites, *TP* true positive, *FP* false positive, *FN* false negative

Table 1 shows that 100% of detection was achieved for the parasites in the low-level complexity and 72% of detecting was reached for parasites in the medium-level complexity. However, in the case of high-level complexity, 0% of parasites were detected. Finally, in the 50 images, a total of 35 objects were automatically detected as parasites, but 10% were false positives.

3.1 Conclusions

In this work, a first prototype of a system for the automatic detection of the *T.* cruzi parasite is presented.clusters. The system includes a commercial optical microscope, an automated mechatronical system to move and control the microscope stage, and the acquisition and processing images systems. The first results show that it is feasible to apply the techniques described in this article, as a method for the detection of the parasite *Trypanosoma cruzi*, which causes Chagas disease. However, it is required to improve the autofocus process during the image acquisition. Regarding the image analysis and processing, it is necessary to improve the segmentation process and incorporate a "spline" algorithm, in order to separate the parasite from erythrocyte clusters.

Concerning the parasite detection problem itself, we believe that if the segmentation process is improved, the detection of parasite in the high and medium levels of complexity will be improved.

It is important to recognize that increasing image magnification ($\times 100$ or $\times 160$) would substantially reduce the difficulties encountered in detecting the parasite [8]. However, the scanning and processing times of the slides would increase considerably. We have already made, in the present research, a significant effort to work with 40-fold magnification.

References

1. World Health Organization. (2019). Chagas disease (American trypanosomiasis) [Online]. Available at: https://www.who.int/news-room/fact-sheets/detail/chagas-disease-(american-trypanosomiasis). Accessed 15 Jan 2019.

2. Prata, P. (2001). Clinical and epidemiological aspects of Chagas disease. *The Lancet. Infectious Diseases, 1*(2), 92–100.
3. Hodo, C. L., Wilkerson, G. K., Birkner, E. C., Gray, S. B., & Hamer, S. A. (2018). Trypanosoma cruzi transmission among captive nonhuman primates, wildlife, and vectors. *EcoHealth, 15*, 426–436.
4. Dumonteil, E. (1999). Update on Chagas disease in Mexico. *Salud Púlica de México, 41*(4), 322–327.
5. Fabrizio, M. C., Schweigmann, N. J., & Bartoloni, J. (2016). Analysis of the transmission of Trypanosoma cruzi infection through hosts and vectors. *Parasitology, 143*(9), 1168–1178.
6. Browne, A. J., Guerra, C. A., Alves, R. V., da Costa, V. M., Wilson, A. L., Pigott, D. M., Hay, S. I., Lindsay, S. W., Golding, N., & Moyes, C. L. (2017). The contemporary distribution of Trypanosoma cruzi infection in humans, alternative hosts and vectors. *Scientific Data, 4*(170050), 1–9.
7. Gomes, G., Almeida, A. B., Rosa, A. C., Araujo, P. F., & Teixeira, A. (2019). American trypanosomiasis and Chagas disease: Sexual transmission. *International Journal of Infectious Diseases, 81*, 81–84.
8. Uc-Cetina, V., Brito-Loeza, C., & Ruiz-Piña, H. (2015). Chagas parasite detection in blood images using AdaBoost. *Computational and Mathematical Methods in Medicine, ID 139681*, 1–13.
9. Rassi, A., & Marin-Neto, J. A. (2010). Chagas disease. *The Lancet, 375*(9723), 1388–1402.
10. Ruiz-Huerta, L., Caballero-Ruiz, A., Vega-Alvarado, L., Heredia-López, F., & Ruiz-Piña, H. (2012). Desarrollo de un sistema automatizado para la detección de parásitos en frotis sanguíneos. In *SOMI XXVII: Congreso de Instrumentacion*. Culiacán, Sinaloa, Mèxico, 29–31 Oct 2012.
11. Billmeyer F. W., & Saltzman M. (2nd Eds.). (1981). *Principles of color technology*. New York: Wiley-Interscience.
12. Vargas-Ordaz, E., Vega-Alvarado, L., Ruiz-Huerta, L., Caballero-Ruiz, A., Heredia-López, F., & Ruiz-Piña, H. (2013). Metodología para la detección en frotis sanguñieos del parásito causante de la enfermedad de Chagas. In *SOMI XXVIII: Congreso de Instrumentacion*. San Francisco de Campeche, Campeche, Mèxico, 28–31 Oct 2013.
13. Bradski, G., & Kaehler, A. (2nd Eds.). (2013). *Learning OpenCV: Computer vision in C++ with the OpenCV library*. Sebastopol: O'Reilly Media, Inc.
14. Pajares, G. (2nd Eds.). (2008). *Visión por computador*. México: Alfaomega.
15. Li, Z., Cheng, Y., Tang, K., Xu, Y., & Zhang, D. (2015). A salt & pepper noise filter based on local and global image information. *Neurocomputing, 159*, 172–185.
16. Gonzalez, R. C., & Woods, R. E. (2nd Eds.). (2002). *Digital image processing*. New Jersey: Prentice-Hall.
17. Russ, J. C. (1st Eds.). (1992) *The image processing Handbook*. Boca Raton: CRC Press.

Monitoring and Evaluating Public Health Interventions

Alfonso Rosales-López and Rosimary Terezinha de Almeida

Abstract Health Technology Assessment (HTA) has become the preferred approach that health systems use for evaluating and monitoring health technologies. Nevertheless, it has mainly focused on pharmaceuticals and medical equipment, while HTAs on public health interventions (PHIs) are rarely performed. The limitations of the traditional methods to evaluate PHIs with a national scope could be one of the reasons for the lack of studies. This situation suggests the need to propose new approaches for evaluating this type of technology. The chapter proposes the use of intervention analysis on time series, using the Box and Tiao approach, as a method for HTA on PHI. Additionally, to illustrate the advantages of the method, a case study is presented in which it is used to assess the impact that the establishment of the National Information System on Breast Cancer, in June 2009, has had on the mortality rates in the five regions of Brazil.

Keywords Health technology assessment · Interrupted time series analysis · Public health intervention · National health programs · Breast neoplasms

1 Health Technology Assessment

Today, more than ever, health technologies are recognized as essential for a sustainable health system. The World Health Organization (WHO) states that equity in public health depends on access to essential, safe, high-quality, affordable, and effective health technologies [44]. However, the procurement, selection, and

A. Rosales-López (✉)
Gerencia de Infraestructura y Tecnología, Caja Costarricense de Seguro Social, San José, Costa Rica

R. T. de Almeida
Programa de Engenharia Biomédica, COPPE, Universidade Federal do Rio de Janeiro, Rio de Janeiro, Brazil

© Springer Nature Switzerland AG 2020   73
M. R. Ortiz-Posadas (ed.), *Pattern Recognition Techniques Applied to Biomedical Problems*, STEAM-H: Science, Technology, Engineering, Agriculture, Mathematics & Health, https://doi.org/10.1007/978-3-030-38021-2_4

incorporation of new technologies into a health system is a process that may vary according to the context in which it is going to be used.

Countries with limited resources require, above all, healthcare policies, practices, and decisions that maximize the positive impact of health technologies on population health while maximizing the cost-benefit of providing them. To that end, the Health Technology Assessment (HTA) has become a fundamental tool to be used to inform the decision-making process concerning the introduction of new technologies to a health system.

According to the International Network of Agencies for Health Technology Assessment [16], HTA is defined as a multidisciplinary process which uses explicit methods to assess the value of a health technology throughout its life cycle. The purpose is to inform decision-making at multiple levels to promote a sustainable and equitable health system. The process is comparative, systematic, unbiased, and transparent and involves multiple stakeholders.

The first studies of HTA began in the early 1970s in the United States, and from that moment, it rapidly spread around the globe [17]. Today, HTA is essential for decision-making in most countries, and several agencies support the advancement of HTA at the global scale. Furthermore, it is common for these agencies to work in networks to share knowledge and experiences, including the INAHTA, the Health Technology Assessment International (HTAi), the European Network for Health Technology Assessment (EUnetHTA), and the HTA Network for the Americas (RedETSA).

In the region of the Americas, the Pan American Health Organization (PAHO, WHO for the Americas) began to promote HTA in 1983, supporting regional meetings, consultations, and the realization of specific assessments. In 1998, PAHO published a Regional Strategy for HTA, and in 2000, PAHO redefined the approach to health technology through interaction with the countries in the region and prioritized the strengthening of the HTA program [2]. Currently, the PAHO promotes the realization of HTA by an integrated approach. This includes the analysis of aspects beyond a traditional assessment, such as the following: the analysis of the context where the technology is going to be used, the health system, different stages of the evaluation, the selection of best therapeutic strategies, the decision-making process, training, regulation, the monitoring of the entire process, and the evaluation of all strategies implemented [30].

Typically, an HTA includes both qualitative and quantitative methods to evaluate different attributes of the technology, such as safety, clinical effectiveness, costs and economic evaluation, and ethical analysis. It also considers the analysis of aspects related to organization, the patients, society, and the regulatory system, among others [16]. For these reasons, HTA has become the preferred approach for evaluating and monitoring health technologies.

While HTA has been mainly focused on pharmaceuticals and medical equipment, HTAs on public health interventions are rarely conducted [10]. In 2010, a survey in five countries showed that only 5% of HTAs were focused on public health interventions [18]. This situation exposes the need to intensify the assessment and monitoring of this type of health technology.

2 Public Health Interventions

Public health interventions (PHIs) are defined as a set of actions intended to promote or protect the health or to prevent ill health in communities or populations [37]. These actions may include policy, regulatory initiatives, single-strategy projects, or multicomponent programs, among others. PHIs comprise multiple interacting components at different organization levels, which increase the difficulty of their implementation [9].

A PHI is considered to be a complex health technology, since the intervention may have limited impact, either because of weaknesses in its design or because it is not properly implemented [35]. In addition, the design depends on the social, organizational, and political settings, since they all influence the effectiveness of the intervention [37].

In that sense, the PHIs are context-dependent in terms of the effects that an intervention may have in different settings even if its implementation does not vary. Also, the health problem targeted by the PHI may differ from one context to another [39]. That is the case of a PHI with national scope, such as a national breast cancer screening program (BCSP).

It is crucial to understand the intervention's design and context for interpreting its outcomes and to generalize them beyond it. Moore et al. [27] suggested that even when an intervention itself is relatively simple, its interaction with its context may still be highly complex. The HTA of a PHI should consider what type of intervention was implemented and the process of how it was done, what the intervention's direct effect is, what the indirect ones could be, and what the objectives at each stage of the intervention are [35].

According to some authors, HTA on PHI poses some challenges that need to be considered. Firstly, a PHI tends to be highly complex due to the varying intervention components, participants, contextual factors, and multiple causal pathways [25]. So, it is crucial to explore the mechanisms through which the intervention brings about change [14]. Besides that, the evaluation is often needed for political reasons, either without randomization, or for a whole population and so without any control [3]. Finally, the evaluation of the effectiveness of the intervention should be matched to its stage of development, and it also should be designed to detect all the side effects and to encapsulate stakeholders' interest [37].

The limitations of the traditional methods for intervention assessment could be the reason for the lack of studies. The randomized controlled trials are considered the gold standard for the evaluation of health interventions. In the field of public health, they are not often available as they are usually difficult to conduct [32], especially for health policies and programs targeted at the population level [3]. Besides that, effect sizes do not provide policy-makers with information on how an intervention might be replicated in their specific context or whether trial outcomes can be reproduced [27].

Alternatively, cohort and case-control studies represent the traditional epidemiological approach to provide important evidence about a disease etiology, but they are less useful for intervention studies, due to limitations such as confusion owing to group differences [41]. In cases of PHI with a national scope, there is no control population to carry out this type of study [36, 38, 42].

Mathes et al. [24] reviewed the existing HTA guidelines for PHI and concluded that methods to evaluate public health interventions are not sufficiently developed or have not been adequately evaluated. Furthermore, Pichon-Riviere et al. [31] concluded that if HTA is not produced and used correctly, it runs the risk of generating inefficient resource allocation by granting coverage to interventions offering little or no benefit, preventing or delaying patient access to useful health technologies, and sending wrong messages to technology producers. This suggests that there is still a need to develop methods and new approaches for evaluating this type of technology.

Recent studies have endorsed to use real-world data (RWD) as an approach to inform effectiveness estimates of novel or existing health technologies, thereby supporting evidence [1, 20]. RWD can be derived from both public and private healthcare databases.

In that sense, Ramsay et al. [34] recommended the use of time series regression techniques, especially when randomized controlled trials are not feasible and it is necessary to evaluate changes in a series after an intervention. This type of study design is increasingly being used for the evaluation of PHI, since it is particularly suited to interventions introduced at a population level over a clearly defined time period [3, 36, 41, 43].

The method of intervention analysis in time series developed by Box and Tiao [5] has proven useful in the assessment of a PHI such as a screening program. A breast cancer screening program (BCSP), for instance, has a national scope, making it difficult to carry out a clinical trial. Its effect on the main outcome (mortality from breast cancer) is indirect, which demands monitoring over time. Besides that, a BCSP is also subject to different interventions.

3   Intervention Analysis on Time Series

A time series is defined as a number of observations taken sequentially in time [4]. In its analysis, the consecutive observations are dependent on each other, and there is interest in modeling this dependency. If each observation is denoted by Z_t (where Z represents the value of the variable at time t), the temporal series with N observations can be defined by the equation:

$$Z(t) = \{Z_1, Z_2, Z_3, \ldots, Z_N\} \tag{1}$$

However, the stochastic process from which the time series is sampled is sometimes affected by special events or conditions such as the implementation of new policies, procedures, programs, and similar events. These events are known as interventions I_t and demand an appropriate approach to analyze the disturbance caused by them in the series, as well as the building of a model to describe the series before and after the occurrence of the intervention [4].

There are two useful types of functions to represent the great majority of interventions in a time series. One is represented by the step function $S_t^{(T)}$, which is used to represent an intervention that should remain active after time T, where T is the moment of the implementation of the intervention, given by the following:

$$S_t^{(T)} = \begin{cases} 0, t < T \\ 1, t \geq T \end{cases} \tag{2}$$

The other type of function is used for interventions that occurred only at time T. Then, its effect is temporary and will disappear over time. It is known as a pulse function, given by the following:

$$P_t^{(T)} = \begin{cases} 0, t \neq T \\ 1, t = T \end{cases} \tag{3}$$

Also, there is information that helps characterize the possible responses on a time series after a specific intervention either by the step or by the pulse functions [4, 19]. This is presented in the response patterns shown in Fig. 1, for both $S_t^{(T)}$ and $P_t^{(T)}$ input functions, where T represents the instant of implementation and T_R the response time.

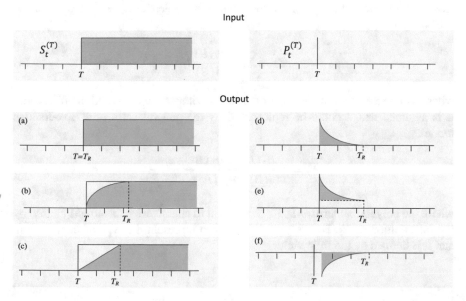

Fig. 1 Interventions' response patterns: (**a–c**) responses to step-type interventions and (**d–f**) responses to pulse-type interventions

The Box-Jenkins models, also known as ARIMA models, are recognized as the most popular approach in time series analysis [7, 23, 26, 28]. Their popularity is due to their adaptive capacity to represent a wide range of processes in a single parsimonious model. In addition, one must take into account the ability to include the presence of interventions into the model [19], which has several advantages that allow reducing possible biases when studying the effects caused by an intervention [34], among these are the following:

- The time series' trend is included in the model. This avoids attributing to the intervention effect of a trend that existed before its occurrence.
- The time series' cyclical and seasonal behavior is also included in the model. As in the previous case, this prevents these behaviors from being attributed to the intervention.
- The models allow the estimation of the response time, that is, the impact time period of the intervention in the time series (T_R in Fig. 1).

3.1 Box and Tiao Approach for Intervention Analysis

Box and Tiao [5] developed a method for intervention analysis on time series using the Box-Jenkins models. In the analysis, it is assumed that an intervention (I_t) occurred at time T of a time series Z_t, and there is an interest to determine whether the intervention caused any change or effect on the series. In the case of a single intervention, the type of the intervention's model for Z_t has the form of the following:

$$Z_t = N_t + \frac{\omega(B)}{\delta(B)} I_t \tag{4}$$

where N_t represents the underlying time series without the intervention effects, and it is assumed that it could be represented by the general form of a Box-Jenkins model:

$$N_t = \frac{\theta_q(B) \cdot \Theta_Q(B^s) \cdot a_t}{\varnothing_p(B) \cdot \Phi_P(B^s) \cdot \nabla^d \nabla_s^D} \tag{5}$$

where the polynomial term $\omega(B)\delta^{-1}(B)$ is used to characterize the effects caused by the intervention on Z_t. For its part, I_t is used to represent the type of intervention, and it is defined by the following:

$$I_t = \begin{cases} 0, t < T \\ 1, t \geq T \end{cases} \tag{6}$$

An advantage of the Box and Tiao approach is its capacity to represent in a single model the effects caused by several interventions on the same time series. To illustrate this situation, Eq. (7) presents an intervention model for a time series Z_t that has been exposed to two interventions I_{1t} and I_{2t} that occurred, respectively, at time T_1 and T_2:

$$Z_t = N_t + \frac{\omega_1(B)}{\delta_1(B)} I_{1t} + \frac{\omega_2(B)}{\delta_2(B)} I_{2t} \tag{7}$$

In the model building, Box and Tiao proposed a process that can be described in four steps: *(i)* preparation of the time series, *(ii)* pre-intervention analysis, *(iii)* the analysis of the intervention's effects, and *(iv)* intervention model building. Each step is described below:

(i) *Preparation of the time series* – The objective is the segmentation of the series to carry out the subsequent pre- and post-analyses. The Z_t series must be segmented at the time of implementation of I_t, where $t = T$. Thus, the first segment Z_{1t} represents the basal series without the intervention's effects and the second segment Z_{2t} the series influenced by the intervention.

$$Z_t = Z_{1t} + Z_{2t} \tag{8}$$

(ii) *Pre-intervention analysis* – The first segment, Z_{1t}, is fitted to a Box-Jenkins model according to its general form:

$$Z_{1t} = \frac{\theta_q(B) \cdot \Theta_Q(B^s) \cdot a_t}{\varnothing_p(B) \cdot \Phi_P(B^s) \cdot \nabla^d \nabla_s^D} \tag{9}$$

(iii) *Analysis of the intervention's effects* – The objective is to identify the effects caused by the intervention. To do so, the model fitted to Z_{1t} is used to forecast the period comprising the length of the second segment Z_{2t}. The intervention's effects are evaluated in the residual series Z_{Re}, which results from the difference between the observed values of Z_{2t} and the forecast values.

The analysis of the residual series aims to discard what represents a white noise process. If Z_{Re} is fitted to a Box-Jenkins model:

$$Z_{Re} = \frac{\omega(B)}{\delta(B)} I_t \tag{10}$$

where $I_t = 0$ for $t < T$.

(iv) *Global intervention model* – The global intervention model (GIM) for the series Z_t results from the sum of the pre-intervention model and the model fitted for the Z_{Re}, which is defined by the equation:

$$GIM : Z_t = \frac{\theta_q(B) \cdot \Theta_Q(B^s) \cdot a_t}{\varnothing_p(B) \cdot \Phi_P(B^s) \cdot \nabla^d \nabla_s^D} + \frac{\omega(B)}{\delta(B)} I_t \tag{11}$$

4 Case Study: Assessing an Action of the Brazilian Breast Cancer Screening Program

A breast cancer screening program (BCSP) is a public health intervention with national scope, making it difficult to carry out an HTA by the traditional methods. Its effect on the main outcome (being the reduction in the mortality from breast cancer) is indirect, which requires monitoring over time. Besides that, a BCSP is also subject to different interventions.

Rosales-López et al. [36] used an approach of intervention analysis to identify the effect that the establishment of the National Information System on Breast Cancer (SISMAMA in Portuguese) in June 2009 caused on the Brazilian mortality rates from breast cancer. The results showed a way to evaluate an action of the BCSP, considering that it led to the ability to quantify the effects and the response time of the intervention (SISMAMA).

Considering that Brazil is known as a country with extremes in income and many social inequalities [40], it can be expected that a PHI would exhibit different results among its five regions, although the PHI was conceived to be applied equally everywhere in the country. Under this situation, it is necessary to apply different analytical approaches for planning purposes. This section presents the use of the intervention analysis approach to assess the impact that the establishment of the SISMAMA has had on the mortality rates from breast cancer in each of the five regions of Brazil.

4.1 Methods

The data of mortality from breast cancer in Brazil and in its five geographical regions were obtained from the Brazilian Mortality Information System [6], where the mortality monthly data since January of 1996 is available. A time series of female mortality rate from breast cancer was built using the number of monthly deaths and the corresponding monthly female population [15], starting in January 1996 and ending in March 2016. All data is publicly available.

Each series was divided into three periods: the pre-intervention period from January 1996 to June 2009, to build a model of the series before the establishment of the intervention; the post-intervention period from July 2009 to December 2014, to analyze the effect of the intervention on the mortality; and the validation period from January 2015 to March 2016, to test the predictive accuracy of the models. All statistical analyses were performed using the *stats* and *forecast* packages of the **R** Statistical Software, version 2.14.0 [33].

4.1.1 Pre-intervention Analysis

According to the method and criteria described by Rosales-López et al. [36], the pre-intervention period of each series was fitted to a Box-Jenkins model, called pre-intervention model (PIM). The mean absolute percentage error (MAPE) was calculated to assess the fitting of the models.

4.1.2 Analysis of the Intervention's Effect

The intervention's effect on the mortality rates was assessed on the residue series, which provided the difference between the observed data post-intervention and the forecast values of the PIM. The Cramér-von Mises statistic was used to discard this residue as a white noise process ($p < 0.05$). When this occurred, it represented the intervention's effect, and the following approach was carried out.

Approach 1 The autocorrelation function (ACF) was used to verify if the residue series was autocorrelated and followed an ARMA process. Then, the series was fitted to an intervention effect model (IEM), which together with the PIM was used to build the global intervention model (GIM).

Then, the GIM was used to estimate the intervention's response time. This is the period that the impact of the intervention takes to be reflected in the time series. For its calculation, the residual series was divided into 11 segments with lengths of multiples of 6 months, varying from 6 up to 66 months. One by one, the segments were used by the GIM to estimate the mortality rates in the post-intervention period. The MAPE was calculated to assess each estimation. These values were plotted sequentially considering the segments' lengths, and the point that presented an abrupt fall was considered as the response time.

Afterward, the post-intervention period was divided in two phases of analysis: (1) A "changing phase" that started with the establishment of the intervention and went on over the response time, and the MAPE was used to assess the fitting of the PIM and GIM to the mortality rates from this phase. (2) A "steady phase" that started once the response time ended and continued until December 2014. The mortality series of this period was fitted to a post-response model (PRM), and the MAPE was used to assess the corresponding fitting of the PIM, GIM, and PRM to this phase.

Finally, aiming to use a model for planning purposes, each model was used to forecast the mortality rates from January 2015 to March 2016. The corresponding predictive accuracy was tested by comparing the results with the observed values for the same period, and the MAPE was used to measure it. Considering the variability within the data of each series, the coefficient of variation was estimated to test the instability of this validation period.

Approach 2 When no intervention effect was identified, the pre- and post-intervention periods were joined and fitted to a non-effect model (NEM). Then, to test the predictive accuracy of the PIM and NEM, the previous strategy was carried out.

Table 1 Pre-intervention models for the period January 1996 to June 2009 in Brazil and in its five regions

Region	PIM	MAPE
South	$(0,1,1)\ (1,1,1)_{12}$	5.88
Southeast	$(1,1,1)\ (1,1,2)_{12}$	3.58
Central West	$(1,1,1)\ (1,1,1)_{12}$	13.74
North	$(0,1,1)\ (2,1,2)_{12}$	18.38
Northeast	$(0,1,1)\ (1,1,1)_{12}$	6.57
Brazil	$(0,1,2)\ (1,1,1)_{12}$	2.85

4.2 Results

Each series is composed of 243 observations. All showed an increasing trend since the beginning of their recording. In general terms, the Central West, North, and Northeast regions showed average rates below the national average, and the South and Southeast regions showed rates above the national average.

The pre-intervention periods were fitted to Seasonal Autoregressive Integrated Moving Average (SARIMA) models. The corresponding PIM structures are shown in Table 1, together with the MAPE values that assessed the fitting of the series to each model.

Figure 2 shows the forecast series of the post-intervention period using the PIM and the observed mortality rates of each region. The forecast series (dashed lines) in Brazil and in its South and Southeast regions showed growth at a slower pace than the observed data (solid gray line). This suggested that the intervention had led to an increase in the mortality rate. For each case, the difference between the observed data and the forecast data was calculated, and the Cramér-von Mises test discarded the residue which was considered to correspond to white noise process (p-values of 1.18e-08, 1.43e-08, and 1.18e-08, respectively). In contrast, this test confirmed that the residue in the Central West, North, and Northeast regions correspond to white noise process (p-values of 0.61, 0.84, and 0.61, respectively).

The three series, Brazil and its South and Southeast regions, were treated according to Approach 1. The ACF of each residue series showed significant components (Fig. 3); therefore, each series was fitted to the IEM: ARIMA $(1,1,0)$ for the residue series of Brazil, ARIMA $(1,1,2)$ for the South region, and ARIMA $(1,1,0)$ for the Southeast region. Consequently, the GIM structures are composed of the sum of the PIM with the corresponding IEM, as follows: $[(0,1,2)(1,1,1)12 + (1,1,0)]$ for Brazil, $[(0,1,1)(1,1,1)12 + (1,1,2)]$ for the South region, and $[(1,1,1)(1,1,2)12 + (1,1,0)]$ for the Southeast region.

Figure 4 shows the MAPE values of the post-intervention period that were calculated using the GIMs for each of the 11 segments of the residual series. In the case of Brazil, an abrupt fall of the MAPE was detected when using the segment of 24 months, which indicated the intervention response time. In the South and Southeast regions, the abrupt fall was observed when using segments of 12 and 24 months, respectively. For each case, the PRM structures were $(1,1,2)(1,1,1)12$ for Brazil, $(0,1,1)(1,1,1)12$ for the South region, and $(2,1,1)(1,1,1)12$ for the Southeast region.

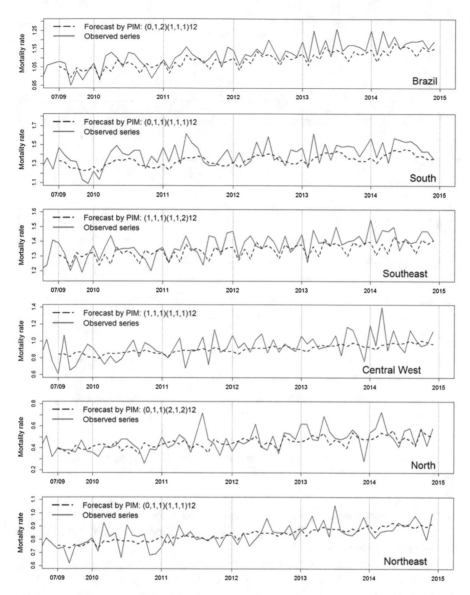

Fig. 2 The post-intervention mortality rate of breast cancer and the series forecast by the PIMs in Brazil and in its five regions

Table 2 shows the results of the analysis in the post-intervention period. During the changing phase, the GIM showed a better fitting than the PIM, meaning that the GIM was able to incorporate the changes due to the intervention. Once the

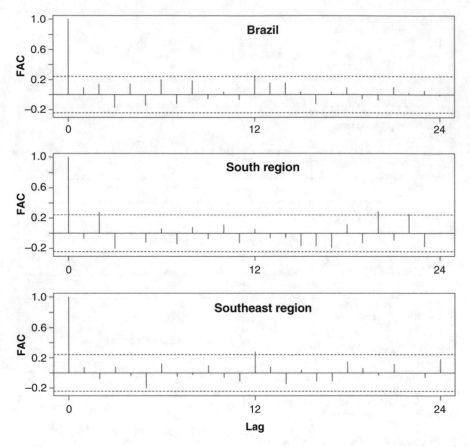

Fig. 3 Correlograms of the ACF on the residue series in Brazil and in the South and Southeast regions

intervention response time passed, in the steady phase, the PRM showed the best fit (even better than the GIM). This suggests that most of the changes occurred during the estimated response time.

The results from the models' predictive accuracy are shown in Table 3. For the Brazilian series, it was shown that the predictions were improved using the PRM: using this model, it was possible to forecast the mortality rates with a 2% error. This is consistent with the results of the assessment of the models' fit. In contrast, the best predictions for the South and Southeast series were obtained using the GIM, and not the PRM. The coefficient of variation indicates greater instability of these series than the Brazil series; therefore, the results are coherent.

The series of the Central West, North, and Northeast regions did not show any effect after the intervention was established, so they were treated according to Approach 2. The joined pre- and post-intervention periods formed a new series that is composed of 228 observations. These series were fitted to SARIMA models

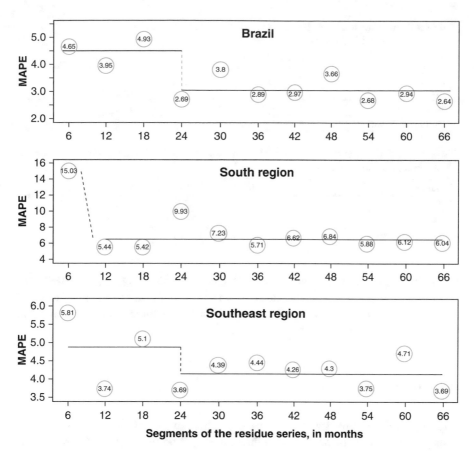

Fig. 4 MAPE of the post-intervention periods estimated by the GIMs and segments of residues in Brazil and in its South and Southeast regions

Table 2 MAPE values to assess the fit of the models in the post-intervention period in Brazil and in its South and Southeast regions

Phase	Model	Brazil	South region	Southeast region
Changing phase	PIM	3.38	6.92	3.64
	GIM	3.26	6.25	3.39
Steady phase	PIM	3.42	6.68	4.27
	GIM	2.28	5.26	3.85
	PRM	1.19	3.54	2.10

and the corresponding NEM structures were (1,1,1)(1,1,1)12 for the Central West region, (3,1,1)(1,1,2)12 for the North region, and (0,1,1)(2,1,2)12 for the Northeast region.

The predictive accuracy of the PIM and the NEM was tested, and the results are presented in Table 4. For the Central West and North series, the NEM improved the

Table 3 MAPE for the forecast series from January 2015 to March 2016 for Brazil and its South and Southeast regions

Model	Brazil	South region	Southeast region
PIM	5.22	7.15	5.51
GIM	3.54	5.30	4.42
PRM	1.98	6.08	4.81
Coefficient of variation	4.33	7.39	5.85

Table 4 Predictive accuracy of the models from January 2015 to March 2016 in the Central West, North, and Northeast regions

Model	Central West region	North region	Northeast region
PIM	10.03	18.73	5.74
NEM	9.33	15.65	5.93
Coefficient of variation	11.37	12.86	8.02

predictions of the mortality rates. However, for the Northeast region, the PIM result was more accurate, but its prediction error was not very distant from the NEM (5.74 vs. 5.93). The coefficient of variation indicates that these three regions present more instability than the previous three.

4.3 Discussion

The approach presented above allows the evaluation of an action of the BCSP considering the effects manifested in the mortality time series of the five regions of Brazil. After the establishment of the SISMAMA, the South and Southeast regions showed an increase in their mortality rates, while the other three regions (Central West, North, and Northeast) did not show any significant change. Furthermore, these results suggest that the changes perceived in the national series (also an increase) do not reflect a generalized result for the entire country but reflect the high influence from the data from the South and Southeast regions, which concentrate approximately 55% of the population [15].

As discussed in the previous work [36], it was expected that the BCSP would reduce mortality, and yet, the identified effect was of growth in the mortality rates. However, it must be taken into consideration that the reduction of mortality is a medium- to long-term effect [45], and in the initial phase of a screening program, an increase in the identification of new cases of cancer is to be expected, especially among women being screened for the first time [21]. In countries such as Canada, Sweden, Norway, and England, where a reduction in the mortality from breast cancer has been observed, the screening actions have, on average, been implemented for more than 26 years [8, 22, 29], while the post-intervention period analyzed in this study corresponds only to the first 6 years of the organization of the Brazilian BCSP.

Considering the intervention, the implementation of the SISMAMA can be represented as a step signal, which equals zero before its implementation and one after it. Nevertheless, the length of time that it took to get to the steady phase was different among regions (12 months in the South region and 24 months in the Southeast region). Different aspects explain this situation, including the differences in geographical dimensions, population, healthcare infrastructure and organization, and many other factors that should be considered. The importance of this result is related to the real possibility of monitoring and evaluating PHIs and their diffusion in such a huge country as Brazil.

Aside from the above considerations, other studies have used a descriptive analysis of the mortality trends to characterize urban and rural areas where the mortality has shown patterns of growth, stability, and even decrease [11–13]. Nevertheless, those studies did not consider the PHIs that occurred over the years, including the BCSP's actions, while our approach did. Another positive aspect of our approach is the possibility of building models that are more suited to the context in which they are going to be used.

When the models were used as forecasting tools, the results showed that the model that best fits the data is not necessarily the best to reproduce the mortality rates. For the Brazilian series, the PRM guarantees the lowest prediction error (MAPE of 1.98), whereas for the South and Southeast regions, the GIM generated the lowest prediction error (MAPE of 5.30% and 4.42%, respectively). This situation is justified based on the higher variability observed in the regional series, which was confirmed by the coefficient of variation (7.39 in the South region and 5.85 in the Southeast region) which showed that the instability of the series interferes with the models' forecast capability.

For the remaining regions, the predictive accuracy was tested on the PIM and NEM. For the Northeast region, there was shown to be no significant difference when using either of the two models for forecasting purposes. This can be interpreted as if in the past 6 years (during the ost-intervention period) the mortality from breast cancer had not been affected by the intervention. The forecasts of the Central West and North regions were improved using the NEM, which suggests that during the same period the mortality rates changed, not necessarily due to the implementation of the SISMAMA.

Although the approach reached its objective, there were some limitations. Firstly, one limitation is the influence that underreporting has on the data, which is related to the difficulty of determining the exact number of deaths caused by breast cancer and the exact number of women per region. Secondly, the use of secondary data is associated with the quality of information contained in the databases that were used to build the mortality rates [13]. Thirdly, this work is limited to the analysis of only one action of the BCSP and, as has been pointed out in the literature, the evolution of breast cancer mortality results from a combination of different risk factors and the effects caused by every action implemented by healthcare systems. Nevertheless, the strength of the approach is precisely its capacity to represent a wide range of processes and interventions in a single parsimonious model.

4.4   Conclusion

This case study demonstrated how intervention analysis on time series results in a useful tool, not only to evaluate a PHI but also to monitor it, especially as this tool was able to identify how an action that was supposed to impact the entire country equally revealed different effects among the regions. This makes this approach a useful assessment tool to provide reliable information for decision-makers.

References

1. Adeagbo, C. U., Rattanavipapong, W., Guinness, L., & Teerawattananon, Y. (2018). The Development of the Guide to Economic Analysis and Research (GEAR) online resource for low-and middle-income countries' health economics practitioners: A commentary. *Value in Health, 21*(5), 569–572.
2. Banta, D., & Jonsson, E. (2009). History of HTA: Introduction. *International Journal of Technology Assessment in Health Care, 25*, 1–6.
3. Bernal, J. L., Cummins, S., & Gasparrini, A. (2017). Interrupted time series regression for the evaluation of public health interventions: A tutorial. *International Journal of Epidemiology, 46*, 348–355.
4. Box, G., Jenkins, G., Reinsel, G., & Ljung, G. (2015). *Time series analysis: Forecasting and control* (5th ed.). Hoboken: Wiley.
5. Box, G. E., & Tiao, G. (1975). Intervention analysis with applications to economic and environmental problems. *Journal of the American Statistical Association, 70*, 70–79.
6. BRASIL. (2015). *Sistema de Informação sobre Mortalidade (SIM) – Informações de Saúde (TABNET) [WWW Document]*. Minist. Saúde – Portal Saúde. URL http://www2.datasus.gov.br/DATASUS/index.php?area=0205&VObj=. Accessed 13 Apr 2016.
7. Chandran, A., Pérez-Núñez, R., Bachani, A. M., Híjar, M., Salinas-Rodríguez, A., & Hyder, A. A. (2014). Early impact of a national multi-faceted road safety intervention program in Mexico: Results of a time-series analysis. *PLoS One, 9*, e87482.
8. Coldman, A., Phillips, N., Warren, L., & Kan, L. (2007). Breast cancer mortality after screening mammography in British Columbia women. *International Journal of Cancer, 120*, 1076–1080.
9. Craig, P., Dieppe, P., Macintyre, S., Michie, S., Nazareth, I., & Petticrew, M. (2008). Developing and evaluating complex interventions: The new medical research council guidance. *BMJ, 337*, a1655.
10. Draborg, E., Gyrd-Hansen, D., Poulsen, P. B., & Horder, M. (2005). International comparison of the definition and the practical application of health technology assessment. *International Journal of Technology Assessment in Health Care, 21*, 89–95.
11. Felix, J. D., Castro, D. S., Amorim, M. H., & Zandonade, E. (2011). Breast cancer mortality trends among women in the state of Espirito Santo between 1980 and 2007. *Revista Brasileira de Cancerologia, 57*, 159–166.
12. Freitas, R., Gonzaga, C. M., Freitas, N. M., Martins, E., & Dardes, R. M. (2012). Disparities in female breast cancer mortality rates in Brazil between 1980 and 2009. *Clinics, 67*, 731–737.
13. Girianelli, V. R., Gamarra, C. J., & Azevedo, G. (2014). Disparities in cervical and breast cancer mortality in Brazil. *Revista de Saúde Pública, 48*, 459–467.
14. Grant, A., Treweek, S., Dreischulte, T., Foy, R., & Guthrie, B. (2013). Process evaluations for cluster-randomised trials of complex interventions: A proposed framework for design and reporting. *Trials, 14*(1), 15.
15. IBGE. (2015). *Instituto Brasileiro de Geografia e Estatística [WWW Document]*. Minist. Planej. URL http://www.ibge.gov.br/home/. Accessed 12 Apr 2016.

16. INAHTA. (2019). *HTA glossary – the international network of agencies for health technology assessment [WWW document]*. URL http://htaglossary.net/HomePage. Accessed 14 Jan 2019.
17. Jonsson, E., & Banta, D. (1999). Management of health technologies: An international view. *BMJ, 319*(7229), 1293.
18. Lavis, J. N., Wilson, M. G., Grimshaw, J. M., Haynes, R. B., Ouimet, M., Raina, P., Gruen, R. L., & Graham, I. D. (2010). Supporting the use of health technology assessments in policy making about health systems. *International Journal of Technology Assessment in Health Care, 26*, 405–414.
19. Liu, L.-M., Hudak, G. B., Box, G. E., Muller, M. E., & Tiao, G. C. (1992). *Forecasting and time series analysis using the SCA statistical system* (1st ed.). DeKalb: Scientific Computing Associates.
20. Makady, A., van Veelen, A., Jonsson, P., Moseley, O., D'Andon, A., de Boer, A., et al. (2018). Using real-world data in health technology assessment (HTA) practice: A comparative study of five HTA agencies. *PharmacoEconomics, 36*(3), 359–368.
21. Malmgren, J. A., Parikh, J., Atwood, M. K., & Kaplan, H. G. (2012). Impact of mammography detection on the course of breast cancer in women aged 40–49 years. *Radiology, 262*, 797–806.
22. Massat, N. J., Dibden, A., Parmar, D., Cuzick, J., Sasieni, P. D., & Duffy, S. W. (2016). Impact of screening on breast cancer mortality: The UK program 20 years on. *Cancer Epidemiology, Biomarkers & Prevention, 25*, 455–462.
23. Masukawa, M. L. T., Moriwaki, A. M., Uchimura, N. S., Souza, E. M., & Uchimura, T. T. (2014). Intervention analysis of introduction of rotavirus vaccine on hospital admissions rates due to acute diarrhea. *Cadernos de Saúde Pública, 30*, 2101–2111.
24. Mathes, T., Antoine, S.-L., Prengel, P., Bühn, S., Polus, S., & Pieper, D. (2017). Health technology assessment of public health interventions: A synthesis of methodological guidance. *International Journal of Technology Assessment in Health Care, 33*, 135–146.
25. Mathes, T., Willms, G., Polus, S., Stegbauer, C., Messer, M., Klingler, C., Ehrenreich, H., Niebuhr, D., Marckmann, G., Gerhardus, A., & Pieper, D. (2018). Health technology assessment of public health interventions: An analysis of characteristics and comparison of methods-study protocol. *Systematic Reviews, 7*, 79.
26. Mellou, K., Sideroglou, T., Papaevangelou, V., Katsiaflaka, A., Bitsolas, N., Verykouki, E., Triantafillou, E., Baka, A., Georgakopoulou, T., & Hadjichristodoulou, C. (2015). Considerations on the current universal vaccination policy against hepatitis a in Greece after recent outbreaks. *PLoS One, 10*, e0116939.
27. Moore, G. F., Audrey, S., Barker, M., Bond, L., Bonell, C., Hardeman, W., Moore, L., O'Cathain, A., Tinati, T., & Wight, D. (2015). Process evaluation of complex interventions: Medical Research Council guidance. *British Medical Journal, 350*, h1258.
28. Nilson, F., Bonander, C., & Andersson, R. (2015). The effect of the transition from the ninth to the tenth revision of the international classification of diseases on external cause registration of injury morbidity in Sweden. *Injury Prevention, 21*, 189–194.
29. Olsen, A. H., Lynge, E., Njor, S. H., Kumle, M., Waaseth, M., Braaten, T., & Lund, E. (2013). Breast cancer mortality in Norway after the introduction of mammography screening. *International Journal of Cancer, 132*, 208–214.
30. PAHO/WHO. (2016). *PAHO's role in health technology assessment in the Americas [WWW document]*. Pan American Health Organization World Health Organization. URL https://www.paho.org/hq/index.php?option=com_content&view=article&id=11581:pahos-role-in-health-technology-assessment-in-the-americas&Itemid=41685&lang=en. Accessed 14 Feb 2019.
31. Pichon-Riviere, A., Soto, N. C., Augustovski, F. A., Martí, S. G., & Sampietro-Colom, L. (2018). Health technology assessment for decision making in Latin America: Good practice principles. *International Journal of Technology Assessment in Health Care, 34*(3), 241–247.
32. Petticrew, M., Chalabi, Z., & Jones, D. R. (2011). To RCT or not to RCT: Deciding when 'more evidence is needed for public health policy and practice. *Journal of Epidemiology and Community Health, 66*(5), 391–396.

33. R Foundation. (2015). *The R project for statistical computing [WWW Document]*. URL https://www.r-project.org/. Accessed 12 Feb 2016.
34. Ramsay, C. R., Matowe, L., Grilli, R., Grimshaw, J. M., & Thomas, R. E. (2003). Interrupted time series designs in health technology assessment: Lessons from two systematic reviews of behavior change strategies. *International Journal of Technology Assessment in Health Care, 19*, 613.
35. Reinsperger, I., Rosian, K., & Winkler, R. (2019). Assessment of public health interventions for decision support: Methods & processes of the evaluation of the Austrian screening programme for pregnant women & children. *Wiener Medizinische Wochenschrift, 169*(11), 263–270.
36. Rosales-López, A., Raposo, L. M., Nobre, F. F., de Almeida, R. T., Rosales-López, A., Raposo, L. M., Nobre, F. F., & de Almeida, R. T. (2018). The use of intervention analysis of the mortality rates from breast cancer in assessing the Brazilian screening programme. *Research on Biomedical Engineering, 34*, 285–290.
37. Rychetnik, L., Frommer, M., Hawe, P., & Shiell, A. (2002). Criteria for evaluating evidence on public health interventions. *Journal of Epidemiology and Community Health, 56*, 119–127.
38. Silva-Illanes, N., & Espinoza, M. (2018). Critical analysis of Markov models used for the economic evaluation of colorectal cancer screening: A systematic review. *Value in Health, 21*(7), 858–873.
39. Shiell, A., Hawe, P., & Gold, L. (2008). Complex interventions or complex systems? Implications for health economic evaluation. *BMJ, 336*, 1281.
40. Solt, F. (2016). The standardized world income inequality database. *Social Science Quarterly, 97*, 1267–1281.
41. Soumerai, S. B., Starr, D., & Majumdar, S. R. (2015). How do you know which health care effectiveness research you can trust? A guide to study Design for the Perplexed. *Preventing Chronic Disease, 12*, E101.
42. Sweeting, M. J., Masconi, K. L., Jones, E., Ulug, P., Glover, M. J., Michaels, J. A., & Thompson, S. G. (2018). Analysis of clinical benefit, harms, and cost-effectiveness of screening women for abdominal aortic aneurysm. *The Lancet, 392*(10146), 487–495.
43. Wagner, A. K., Soumerai, S. B., Zhang, F., & Ross-Degnan, D. (2002). Segmented regression analysis of interrupted time series studies in medication use research. *Journal of Clinical Pharmacy and Therapeutics, 27*, 299–309.
44. WHO. (2017). *WHO list of priority medical devices for cancer management, WHO medical device technical series*. Geneva: World Health Organization.
45. WHO. (2012). *Breast cancer: Prevention and control [WWW Document]*. World Health Organization Programme Projects Cancer. URL http://www.who.int/cancer/detection/breastcancer/en/. Accessed 12 Apr 2016.

Recognition of Nausea Patterns by Multichannel Electrogastrography

Millaray Curilem, Sebastián Ulloa, Mariano Flores, Claudio Zanelli, and Max Chacón

Abstract Nausea is a common set of symptoms related to several underlying physiological causes, usually difficult to identify a priori. Detecting nausea before emesis (vomiting) is particularly important for patients who are still unconscious after surgery, because emesis may cause various life-threatening complications. Electrogastrography (EGG) is the cutaneous measurement of the electrical activity of the stomach sensed by electrodes placed on the abdomen of the patient. As the relationship between nausea and gastric dysrhythmias is not yet well understood, the study of electrogastrograms may generate information to relate these processes. Thus, the aim of this study was to evaluate the possibility of detecting the presence of nausea in chemotherapy patients. The method consists of acquiring signals using multichannel electrogastrograms, isolating the gastric motion information applying independent component analysis, and then processing the resulting signal to discriminate between normal function and nausea. Feature extraction, clustering, and selection yielded a classifier that discriminated between both classes. The performance of the classifiers was compared among different experiments which include the number of channels and the period of nausea observation (pre- and

M. Curilem (✉)
Centro de Física e Ingeniería para la Medicina (CFIM), Universidad de La Frontera, Temuco, Chile
e-mail: millaray.curilem@ufrontera.cl

S. Ulloa
Departamento de Ingeniería Eléctrica, Universidad de La Frontera, Temuco, Chile

M. Flores
Hospital Regional Dr. Hernán Henríquez Aravena, Temuco, Chile
e-mail: mariano.flores@ufrontera.cl

C. Zanelli
Onda Corporation, Sunnyvale, CA, USA
e-mail: et@ondacorp.com

M. Chacón
Departamento de Ingeniería Informática, Universidad de Santiago de Chile, Santiago, Chile
e-mail: max.chacon@usach.cl

© Springer Nature Switzerland AG 2020
M. R. Ortiz-Posadas (ed.), *Pattern Recognition Techniques Applied to Biomedical Problems*, STEAM-H: Science, Technology, Engineering, Agriculture, Mathematics & Health, https://doi.org/10.1007/978-3-030-38021-2_5

post-detection). The best classifier obtained 83.33% of accuracy discriminating 31 control and 29 nausea events, using only 4 EGG channels and 3 features: the dominant power, the dominant frequency, and the relationship between the maximum and the average power of the event.

Keywords Nausea · Multichannel electrogastrography · Independent component analysis · Support vector machines · Chemotherapy patients

1 Introduction

The physiopathology of nausea in humans is still poorly understood, given the number and variety of diagnostic possibilities [35]. This is why the clinical approach is complex. The prevalence of postoperative nausea and vomiting (PONV) can complicate 11–73% of the surgical procedures, with incidence rates of 37–91% [1, 32]. PONV remains one of the most frequent and painful complications after surgery, resulting in pain, bruising, and wound dehiscence, which requires additional resources and may delay patients' recovery [24]. Patients with persistent PONV in the surgery units are at greater risk in the postoperative period, especially during the 24 hours after surgery. PONV management is essential because patients are in recovery and in many cases have not yet fully regained consciousness [25].

Although nausea is very frequent and critical in this type of patients, there is not yet a mechanism that allows this symptom to be identified prematurely [23]. The treatment is usually prophylactic for those patients whose historical background and experience with previous surgeries, statistically, cause postoperative nausea more frequently.

Several works have identified factors that influence the occurrence of PONV. These factors are mainly sex, age, the history of previous occurrences, history of smoking, the type of surgery, and obesity, among others. Based on all or some of these factors, several studies have been carried out to construct predictors of PONV. Sinclair et al. [32] used an analysis based on logistic regression to analyze a survey applied to 17,638 pre- and postoperative patients whose objective was to determine the factors that most influence the incidence of PONV. This analysis allowed obtaining a predictor of PONV that reaches a 78.5% accuracy. Peng et al. [29, 30] analyzed 1086 cases and developed a Bayesian classifier that reached 77% accuracy, and an artificial neural network was trained, reaching an accuracy of 83.3%. Lee et al. [25] describe an analysis based on decision trees and a predictor that was implemented over 1181 cases, reaching 88.1% of accuracy. Hutson et al. [15] performed a study on 50,408 patients who received general anesthesia and were scored as being at high risk for PONV. The results improved the prophylactic measures for high-risk patients. From previous studies, it follows that there is no effective method of early nausea detection, which implies the widespread use of antiemetic drugs when epidemiological information indicates a high probability of occurrence. However, it is interesting to look for other approaches that can more effectively detect early nausea and that may improve the understanding of the related physiological phenomena.

Presumably, gastric motor dysfunction is related to nausea and emesis [11, 20, 21], so the study of stomach motility via its associated electrical behavior should contribute to understanding the mechanisms of these conditions. For this reason, electrogastrography (EGG) which measures the myoelectrical activity of the stomach [37] is expected to generate information to detect nausea states early in unconscious patients.

EGG is the recording of electrical potentials that occur in the muscles of the stomach and intestine [3]. The recording electrodes are placed on the abdomen, providing noninvasive (external) measurements of the gastric electrical slow-wave activity [22]. This electrical activity in turn mainly reflects the contractility of the stomach through changes in the amplitude and frequency of the EGG signals [7]. EGG is a promising tool based on the assumption that there are pathologies that can alter the gastric electrical waves [2]. Many studies have obtained encouraging results identifying certain gastric states, mainly using frequency and power as signal parameters [3, 10]. However, EGG contains signals from various sources, such as cardiac, respiratory, and other artifacts, which makes the signal very noisy, requiring advanced filtering techniques [8]. Moreover, the clinical experience with the use of EGG is affected by the lack of a standardized methodology in terms of the position of the electrodes, recording periods, test meals, analysis software, and normal reference values [31]. For this reason EGG is not yet used massively [37].

Despite these difficulties and controversial use of the EGG records, some studies have found a correlation between episodes of nausea and abnormal myoelectrical activity [11, 17, 28, 34]. This is why the objective of this work was to apply advanced pattern recognition techniques to advance the detection of nausea events through EGG records, in order to support preventive actions and to improve rescue therapies in the postanesthesia care units. The general methodology of this preliminary study is to apply the independent component analysis (ICA) to obtain a single signal from the EGG multichannel records that contains gastric components, eliminating the electrical components generated by other organs. A feature extraction and selection study was also performed analyzing which of the features suggested by the literature improve the discrimination between nausea and control events. Finally, a classifying structure based on support vector machines (SVM) was trained to perform the two-class classification.

2 EGG Dataset

The EGG signals were obtained by an electrogastrography provided by Onda Corporation,[1] a Company specialized in ultrasound measurement instrumentation and services for medical and industrial applications. The sensing device has 10 amplifiers with an impedance of 10 MOhms. Each channel has a bandwidth from

---

[1]Onda Corporation: http://www.ondacorp.com.

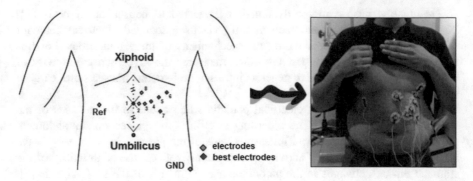

Fig. 1 Position of the electrodes

0.6 to 15 cpm (cycles per minute) (0.01–0.25 Hz) which feeds into a 24-bit digitizer that broadcasts the values over Bluetooth. The device has 10 electrodes that include 8 active channels, the reference, and ground. Conductive gel electrodes are placed following the gastroenterologist instructions, as shown in Fig. 1. Here is the location of the channels used:

- Electrode 1: At the midpoint between the tip of the sternum and the navel.
- Electrode 2: 4 cm to the left (of the patient) of electrode 1.
- Electrode 3: 4 cm to the left of electrode 2, draw a diagonal from electrode 2, going up 30°.
- Electrode 4: 4 cm from electrode 3 along the diagonal.
- Electrode 5: 4 cm from electrode 4 along the diagonal.
- Electrode 6: 4 cm from electrode 6 along the diagonal.
- Electrode 7: Pancreatic zone, without defined position.
- Electrode 8: Located at the intersection point between a vertical vector between electrodes 2 and 3 and a horizontal vector between electrodes 5 and 6.
- GND: In the hip.
- REF: 20 cm from node 1, to the patient's right.

The Bluetooth signals are received by a portable computer for storage in conjunction with operator input related to the patient's observation. From the point of view of safety and comfort, the device is light, electrically isolated from the room, and electrically limited to deliver a worst-case maximum current of 1 μA between electrodes.

The EGG records were obtained from patients of the Hematology Unit of the Regional Hospital, Dr. Hernán Henríquez Araneda, Temuco, Chile. These patients received standard antiemetic as they were susceptible to nausea because of the chemotherapy treatment [6]. The study was authorized by the Ethics Committee of the Araucanía Sur Health Service, as well as by the Hospital. All the patients were submitted to the same acquisition protocol, which included position of the electrodes, the duration of the recordings, etc. They were told to be as calm and

Channel

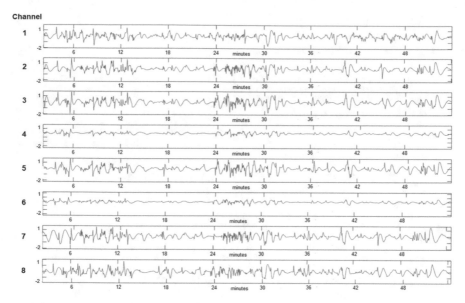

Fig. 2 EGG signals recorded by the eight electrodes

silent as possible and to mention nausea episodes as early as noticed. A trained nurse annotated into the record all events (nausea, movement, talking, etc.). For efficiency, a code was used for the most common events, such as quiet patient (0), patient with nausea (1), vomiting (2), and speech (3). The comments were synchronically recorded with the EGG samples using software installed in the receiver computer. The software generates a binary file containing the sequential number samples followed by the data of 10 channels and a text file containing the number of the sample and the nurse indications. Figure 2 shows the EGG signals obtained from the eight electrodes in 1 hour of recording.

Only nausea events were considered. The duration of the nausea events was established in 2 minutes (334 samples) according to the empirical observations and the medical indications. To define the control events, a careful segmentation was performed to the EGG extracting episodes of calm (no nurse observations). Control episodes have the same duration as the nausea. Figure 3 presents the signals and the spectral content of the EGG signals.

Thirteen EGG were obtained from patients who agreed to participate in the study, and twenty-nine nausea events and thirty-one control were observed in all. Two ways define the beginning of a nausea episode: at the exact time when the patient signaled nausea and 30 seconds before this sample (pre-nausea), to study the onset of the nausea signal. In both cases, the observation of the event spanned 120 seconds.

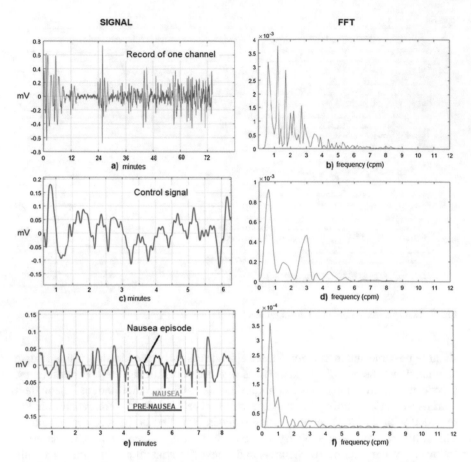

Fig. 3 EGG signals (left) and the amplitude of their spectral content (right). The whole signal recorded by one channel is presented in figures (**a**), a control signal is presented in figures (**c**), and a signal that contains a nausea episode is presented in figures (**e**). The Fast Fourier Transform (FFT) are presented in figures (**b**, **d**, and **f**). Figure (**e**) shows the nausea and pre-nausea segmentation in respect of the beginning of the nausea episode. Figure (**f**) shows the Fourier transform just for the pre-nausea episode (red signal)

3 Nausea Discrimination Method

The methodology used in the present work is presented in Fig. 4. The preprocessing step filtered and normalized the signals. Then, the independent component analysis step was in charge of obtaining a single signal that contains the gastric information retrieved by the multichannel EGG analysis. Next, a feature extraction and selection step was applied to each event, to get its descriptors. Finally, the classification step is performed to discriminate between nausea and control events. The classification performance was used to define the best components of the whole process (ICA impact, feature selection, and the best channel subset).

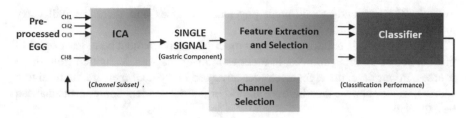

Fig. 4 General structure of the nausea/control discrimination methodology

3.1 Preprocessing Step

Two Butterworth filters were applied to the EGG records: a fourth-order high-pass filter with cutoff frequencies of 0.0083 [Hz] (0.5 cpm) and a seventh-order low-pass filter with a cutoff frequency of 0.15 [Hz] (9 cpm). These ranges are those suggested in the literature [8, 37] and that gave better empirical results.

The signals were processed in a unipolar mode, which implies that the eight channels are obtained from the difference between the electrode measure and the reference. Each digital count corresponds to 1.35×10^{-10} [mV], and the electrogastrography uses 24 bits to code the amplitude, so the center (0 V) is approximately in the value 8,400,000. The expression of Eq. (1) was applied to normalize the EGG:

$$y = \frac{x - 8,400,000}{8,400,000} \tag{1}$$

where x corresponds to the sample value and y to the normalized EGG amplitude.

3.2 Independent Component Analysis (ICA)

Independent component analysis (ICA) belongs to the group of blind source separation methods that allow extracting underlying information components from signals. The term "blind" refers to the fact that these methods can separate the information that comes from unknown sources that were mixed. ICA is based on the assumption that if different signals come from different physical processes, they are statistically independent. ICA takes advantage of the fact that the implication of this assumption can be reversed, giving rise to a new assumption that cannot be guaranteed but that works in practice: if statistically independent signals can be extracted from a signal, then these signals must come from different processes. Consequently, ICA separates the statistically independent components of a signals. If the hypothesis of statistical independence is valid, each of the signal obtained through ICA responds to different physical processes and therefore will carry independent information [18].

The ICA analysis requires that the number of receivers must be at least equal to the number of emitting sources, in order to recover the signals emitted from those sources. In our case, the eight electrodes arranged in the abdomen are the receiving sources, so we suppose that there are at most eight independent components that can be extracted.

$$x = \begin{pmatrix} a & b \\ c & d \end{pmatrix} \begin{pmatrix} s_1^1 & s_1^2 & \ldots & s_1^N \\ s_2^1 & s_2^2 & \ldots & s_2^N \end{pmatrix}$$
$$x = As \quad \Rightarrow \quad s = A^{-1} \cdot x$$

(2)

Equation (2) defines a signal x that is being registered by two receptors. It is assumed that this signal is a linear combination of two independent emitting signals s_1 and s_2. The matrix with the components a, b, c, and d is unknown and is called the matrix of the mixing coefficients. This matrix, multiplied by the original signals, gives rise to the signal received by each receptor. Therefore, the inverse matrix A^{-1} has to be calculated to recover the signals s_1 and s_2 from x. The mixing coefficients of matrix A are obtained by the ICA method, through the theory of maximum likelihood [16].

ICA appears to be an interesting alternative for EGG because there are several sources of bioelectrical signals being recorded in the abdomen. In addition, the registers are multichannel, so it is possible to identify different sources. The separation of these signals in independent signals can improve the acquisition and depuration of the gastric signal, which is the signal of interest, eliminating other sources of electrical activity such as those of other organs or muscles [7].

Here, the FastICA algorithm [16] was applied to obtain the gastric signal from the channels. However, to define which channels give more information, an exhaustive search was made applying ICA to different subsets of channels and comparing the performance of the classifier in each case. This process allowed to determine that channels 4, 5, 6, and 7 (see Fig. 1) improved the result of the classification, yielding better results than those that included all the channels.

3.3 Feature Extraction

The feature extraction process defines which information will be extracted from the events to facilitate the discrimination between the nausea and control classes. As the analysis responds to a medical need, it was considered important that the extracted features have clinical meaning [9].

A set of characteristics extracted from the literature were considered [4, 5, 8, 33], predominantly extracted from the spectral domain. The power and dominant frequency of the spectra are the most used characteristics, as well as the stability of the frequency. Some of these characteristics are presented in the histograms of

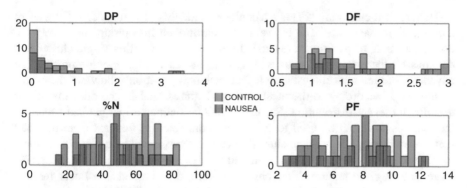

Fig. 5 Histograms of the DP, DF, %N, and PF features according to the control and nausea classes

Fig. 5 for the ICA method. Therefore, the features extracted from each event are as follows:

- Dominant power (DP): Maximum amplitude of the Fourier spectrum.
- Dominant frequency (DF): Frequency associated to the dominant power.
- Power factor (PF): Ratio between the dominant power and the average power. This measures the relationship between the maximum amplitude of the spectrum and all the other spectral components.
- Percentage of normal (%N), bradygastric, and tachygastric waves. These features measure the ratio between the power content of the different frequency bands (normal around 3 cpm, bradygastric under 2 cpm, and tachygastric over 5 cpm) and the total power of the event. It is calculated by dividing the power content of each band by the total power content of the episode.

It is important to note that the features will be extracted from each of the channels in the multichannel processing and from the single signal obtained from the ICA processing. Therefore, in the analysis without ICA, there are 6 features per channel, that is, 48 features, and in the ICA case only 6 features.

In order to work with values in the range 0–1, the features were normalized dividing them by their maximum values.

3.4 Feature Selection

It is important that the set of features that represent the events be appropriate to the classification problem, so that the input space may be divided into separable decision regions associated to the classes. In addition, the number of features has to be reduced to simplify the complexity of the classifier. This is why a feature selection process is needed [13].

Hierarchical clustering [27] is a technique to identify structures or subclasses of the objects in databases. The N events are partitioned into groups iteratively. To do this, a series of partitions are carried out, which start with a single cluster that contains all the events and ends with N clusters, each with a single object. The advantage of hierarchical clustering is that each new partition subdivides the events in subgroups according to the most relevant features, that is, the ones that better describe the event distribution. So, the technique defines a hierarchical order for the features so it can be used to select the more relevant features from the total features extracted. Hierarchical clustering can be represented by a two-dimensional diagram known as a dendrogram, which illustrates the divisions in each successive stage, showing the features hierarchy. Therefore, this method is a linear method for ordering the features, from most to least significant, and can be used as a filter method to select the most significant characteristics that influence the discrimination of the groups [12]. This method was used to select the features for each simulation.

3.5 Classification

The complexity of EGG signal processing leads to the use of more sophisticated techniques for analysis such as artificial neural networks [19] and support vector machines (SVM) [4, 26]. However, SVM has presented better results in previous works [10, 38].

The SVM are a set of supervised learning algorithms that address the problems of classification by transforming the input space into a space of high-dimensional characteristics, where it is possible to find a linear hyperplane of decision that will separate two classes. The functions responsible for the change in dimensionality are known as kernel functions. The Radial Basis Function (RBF) function is the most used kernel function. The hyperplane separator is obtained by means of an optimization algorithm that finds the greatest distance between the points closest to the boundary between the classes. These points are called support vectors. The advantage of using SVM in this application is that they are able to deal with the nonlinearity inherent to the features extracted from the electrogastrograms, like neural networks, but with a smaller number of design parameters to be adjusted [4].

Support vector machine tackles classification problems by nonlinearly mapping input data into high-dimensional feature spaces, wherein a linear decision hyperplane separates two classes, as described in Fig. 6a and b [36].

The decision hyperplane parameters (w, b) are obtained by an optimization algorithm that finds the largest distance (margin) to the nearest training data points of any class, called the support vectors. They define the supporting hyperplanes shown in Fig. 6a. Meanwhile, Fig. 6b shows the soft margin situation, when some points are allowed to cross the supporting hyperplanes. Slack variables ξ are error terms that measure how far a particular point lies on the wrong side of its respective supporting hyperplane. To solve nonlinear problems, it is necessary to transform

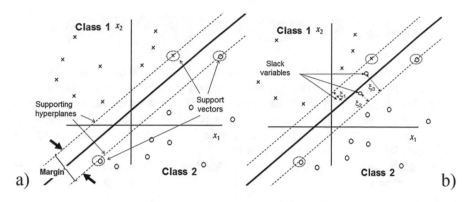

Fig. 6 Linear decision hyperplane of two classes: (**a**) maximum margin and (**b**) soft margin

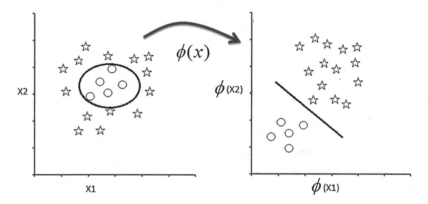

Fig. 7 Effect of mapping the input space into a higher-dimensional feature space, where a linear separation plane is possible

the input space where the data is not linearly separable into a higher-dimensional space called a feature space where the data is linearly separable. The transformation function φ maps the input space into the feature space, as shown in Fig. 7.

These transforming functions called kernels allow computing the inner product between two independent original variables $\vec{x}_i \cdot \vec{x}_j$ in the input space as if it was in the feature space, making it unnecessary to fully evaluate the transformation. Equation (3) shows the most generally used RBF kernel:

$$k\left(\vec{x}_i, \vec{x}_j\right) = \vec{x}_i \cdot \vec{x}_j = \varphi\left(\vec{x}_i\right) \cdot \varphi\left(\vec{x}_j\right) = \frac{e^{-\|\vec{x}_i - \vec{x}_j\|^2}}{2\sigma^2} \tag{3}$$

The optimization problem is then expressed as shown in Eq. (4):

$$\min_{w,b,\xi} \left(\frac{1}{2} w^T w + c \sum_{k=1}^{N} \xi_k \right) s.t. \quad \begin{array}{l} y_k \left[w^T x_k + b \right] \geq 1 - \xi_k \\ \xi_k \geq 0, \quad k = 1, \ldots, N \end{array} \tag{4}$$

where c is a hyperparameter determining the trade-off between the complexity of the model, expressed by $\vec{w}$, and the points that remain in the wrong side of the decision hyperplane, expressed by the slack variables ξ. The solution of this minimization problem for obtaining the weight vectors is found by the standard optimization procedure for a problem with inequality restrictions when applying the conditions of Karush-Kuhn-Tucker to the dual problem.

In classification tasks with RBF kernel, SVM has two hyperparameters to adjust: the RBF parameter σ and the penalty parameter c. These hyperparameters are adjusted in a trial-and-error process called grid search [14].

3.6 Performance Evaluation

The training and validation process of the classifiers was performed through a threefold cross-validation technique. The contingency tables were built, and four statistical indices were calculated to measure the performance of the classifiers: sensitivity (Se) measures the ability of recognizing nausea events, specificity (Sp) measures the ability of recognizing the control events, exactitude (Ex) measures the percentage of events correctly classified, and error (Er) measures the percentage of events incorrectly classified. Equations (5), (6), (7), and (8) show the ratios for each index:

$$Se = \frac{TP}{TP + FN} \tag{5}$$

$$Sp = \frac{TN}{TN + FP} \tag{6}$$

$$Ex = \frac{TP + TN}{n} \tag{7}$$

$$Er = \frac{FP + FN}{n} \tag{8}$$

where true positives (TP) is the number of events correctly classified as being of the positive class, true negatives (TN) is the number of events correctly classified as being of the negative class, false positives (FP) and false negatives (FN) are the number of events classified erroneously, and n is the total number of events. TP, TN, FP, and FN were extracted from the contingency table.

4   Case Study and Results

The simulations were performed on the nausea and the pre-nausea event segmentation. The features were directly extracted from the EGG signals recorded by the electrodes or to the single signal obtained from the ICA decomposition. Finally, the simulations considered all the channels and only the best four channels 4, 5, 6, and 7. This section presents the results obtained from all the simulations that are summarized below.

For the nausea and pre-nausea segmentation, the simulations are as follows:

1. Multichannel all the channels. A classifier was trained with all the features extracted directly from the eight channels.
2. Multichannel channels 4, 5, 6, and 7. A classifier was trained with the features obtained from channels 4, 5, 6, and 7.
3. ICA all the channels. A classifier was trained with the features extracted from the signal obtained through the ICA process applied to the eight channels.
4. ICA channels 4, 5, 6, and 7. A classifier was trained with the features extracted from the signal obtained through the ICA process applied to channels 4, 5, 6, and 7.

4.1   Results for the Feature Selection Process

Table 1 shows the results of the feature selection performed for each simulation considering the nausea segmentation, while Table 2 presents the features selected for the pre-nausea segmentation.

The tables show the significant reduction in the number of features produced by ICA, which is relevant to reduce the complexity of the classifier design. The more selected feature is the Dominant Power followed by the Power factor which indicates that the power of the signals is relevant for the nausea discrimination. Finally it may be observed that even when all the channels are considered, the features extracted from the 4, 5, 6 and 7 channels appear to be relevant.

4.2   Results for the Classification

The comparison of the results was made based on the percentage of classification obtained by the SVM classifiers, trained and validated with the same data folds. Table 3 shows the performance indices for each simulation considering the nausea segmentation, while Table 4 presents the performance indices for the pre-nausea segmentation.

These results show the importance of the segmentation of the nausea events as the indices improved significantly for the pre-nausea segmentation presented in Table 4.

Table 1 Selected features for the nausea segmentation

Simulation	Selected features
Multichannel: all the channels	Power factor channel 1 Percentage of power in the normal band channel 2 Power factor channel 3 Percentage of power in the bradygastric band channel 3 Percentage of power in the normal band channel 3 Percentage of power in the tachygastric band channel 3 Dominant power channel 4 Dominant power channel 5 Percentage of power in the bradygastric band channel 8
Multichannel: channels 4, 5, 6, and 7	Dominant power channel 5 Dominant power channel 6 Power factor channel 6 Power factor channel 5
ICA: all the channels	Dominant power Dominant frequency Power factor
ICA: channels 4, 5, 6, and 7	Dominant power Dominant frequency Percentage of power in the tachygastric band

Table 2 Selected features for the pre-nausea segmentation

Simulation	Selected features
Multichannel: all the channels	Dominant power channel 1 Percentage of power in the tachygastric band channel 2 Dominant power channel 5 Power factor channel 6 Percentage of power in the bradygastric band channel 6 Dominant power channel 7 Percentage of power in the bradygastric band channel 8
Multichannel: channels 4, 5, 6, and 7	Dominant power channel 5 Power factor channel 4 Dominant power channel 6 Power factor channel 6
ICA: all the channels	Power factor Percentage of power in the normal band
ICA: channels 4, 5, 6, and 7	Dominant power Dominant frequency Power factor

The selection of the best channels improved the sensitivity and the specificity indices except for the multichannel simulations Se index of Table 4. The first and forth simulation Se indices for the pre-nausea segmentation are very similar; however, Sp indices improved for the best channel simulations showing the importance of the location of the electrodes to get better classifying performance. It must be underlined that the good results presented for the ICA channels 4, 5, 6, and 7 for the pre-nausea segmentation are reached with only three features.

Table 3 Performance indices for the nausea segmentation

	Performance indices				SVM hyperparameters	
	Se (%)	Sp (%)	Ex (%)	Er (%)	C	Sigma
Multichannel: all the channels	66.29	83.63	75	25	8	2.8284
Multichannel: channels 4, 5, 6, and 7	79.62	61.51	70	30	2	6.498
ICA: all the channels	44.44	77.57	61.66	38.33	5.27	9.84
ICA: channels 4, 5, 6, and 7	48.51	77.57	63.33	36.66	0.0312	8

Table 4 Performance indices for the pre-nausea segmentation

	Performance indices				SVM hyperparameters	
	Se (%)	Sp (%)	Ex (%)	Er (%)	C	Sigma
Multichannel: all the channels	79.62	65.45	71.67	28.33	4	0.7071
Multichannel: channels 4, 5, 6, and 7	75.92	80.60	78.33	21.67	137.187	1.7411
ICA: all the channels	72.22	83.33	78.33	21.67	724.07	4
ICA: channels 4, 5, 6, and 7	82.96	83.63	83.33	16.66	724.07	5.6568

5 Conclusions

Inspired by the work of Peng et al. [29, 30], the present study applied a pattern recognition approach to detect nausea in EGG records. Two kinds of segmentation strategies were applied to define the nausea event, both related to the beginning of the nausea. In the first case (nausea events), the beginning was established at the time (sample) that the patient signaled of having nausea. In the second case (pre-nausea events), it was established 30 seconds before. The control events were segmented from the background EGG signals. Nausea, pre-nausea, and control events span 2 minutes. A classification stage was implemented to discriminate between nausea/pre-nausea events and the controls. In addition to the standard feature extraction and classification processes, an ICA was performed. On the one hand, this technique provided a single signal instead of processing all the EGG channels, thus reducing the complexity of the classifiers, and on the other hand, ICA isolated the gastric information from other sources of electrical activity, yielding a cleaner signal. Nevertheless, the experiments showed that the quality of that signal varied depending on the channels considered, leading to reducing the eight channels to only four.

The results showed that the performance of the classifiers is improved by the pre-nausea segmentation and the four-channel ICA. As expected, the number of features was reduced by the application of ICA. The use of ICA modified the distribution of the events in the feature space, facilitating the discrimination. This is encouraging for the hypothesis that the signal obtained by this method has more information than the independent channels. It is interesting that the electrode location influence the results, as the classification was improved when fewer electrodes were considered.

The results are encouraging, because they show good separation between nausea events from the normal state of a patient (control). For a more encompassing method, however, it would be important to test the method when other signal disturbances are found, for example, when the patient moves or speaks or when there are other gastric disturbances present. For a robust performance, it would be important to test the method with more records to validate the present results and to increase the number of classes to evaluate the method in more real scenarios.

Acknowledgments The authors would like to thank the Research Unit of Universidad de La Frontera for financing this project and to Onda Corporation for providing the equipment necessary for this work.

References

1. AGA. American Gastroenterological Association. (2001). Medical position statement: Nausea and vomiting. *Gastroenterology, 120*, 261–262.
2. Alvarez, W. C. (1922). The electrogastrogram and what it shows. *JAMA, 78*(15), 1116–1118.
3. Babajide, O., & Familoni. (1994). Validity of the cutaneous electrogastrogram. In *Electrography: Principles and applications* (Vol. 1, pp. 103–125). New York: Raven Press.
4. Chacón, M., Curilem, G., Acuña, G., Defilippi, C., Madrid, A. M., & Jara, S. (2009). Detection of patients with functional dyspepsia using wavelet transform applied to their electrogastrograms. *Brazilian Journal of Medical and Biological Research, 42*(12), 1203–1209.
5. Chang, Y. (2005). Electrogastrography: Basic knowledge, recording, processing and its clinical applications. *Journal of Gastroenterology and Hepatology, 20*, 502–516.
6. Chen, B., Hu, S., Liu, B., Zhao, T., Li, B., Liu, Y., et al. (2015). Efficacy and safety of electroacupuncture with different acupoints for chemotherapy-induced nausea and vomiting: Study protocol for a randomized controlled trial. *Trials, 16*(1), 212.
7. Chen, C., Hu, C., Lin, H., & Yi, C. (2006). Clinical utility of electrogastrography and the water load test in patients with upper gastrointestinal symptoms. *Journal of Smooth Muscle Research, 42*(5), 149–157. https://doi.org/10.1540/jsmr.42.149.
8. Chen, J. Z., & McCallum, R. W. (1994). Electrogastrographic parameters and their clinical significance. In *Electrography: Principles and applications* (Vol. 1, pp. 45–72). New York: Raven Press.
9. Chen, J. Z., & McCallum, R. W. (2000). Non-invasive feature based detection of delayed gastric emptyng in humans using neuronal networks. *IEEE Transactions on Biomedical Engineering, 47*, 409–412.
10. Curilem, M., Chacón, M., Acuña, G., Ulloa, S., Defilippi, C., & Madrid, A. M. (2010). Comparison of artificial neural networks and support vector Machines for Feature Selection in Electrogastrography signal processing. *Annual International IEEE EMBS Conference, 2010*, 2774–2777.
11. DiBaise, J. K., Brand, R. E., Lyden, E., Tarantolo, S. R., & Quigley, E. M. M. (2001). Gastric myoelectrical activity and its relationship to the development of nausea and vomiting after intensive chemotherapy and autologous stem cell transplantation. *American Journal of Gastroenterology, 96*(10), 2873–2881.
12. Guyon, I. (2003). Fast an introduction to variable and feature selection. *Journal of Machine Learning Research, 3*, 1157–1182.
13. Guyon, I., Gunn, S., Nikravesh, M., & Zadeh, L. (2006). Feature extraction, foundations and applications. In I. Guyon, S. Gunn, M. Nikravesh, & L. Zadeh (Eds.), *Series studies in fuzziness and soft computing*. Berlin, Germany: Physica-Verlag, Springer.

14. Hsu, C. W., Chang, C. C., & Lin, C. J. (2003). *A practical guide to support vector classification.* Taipei: Department of Computer Science, National Taiwan University.

15. Hutson, L. R., Ragsdale, S. A., & Vacula, B. B. (2019). Relation of improved postoperative nausea/vomiting quality metric to physician incentive pay. *Baylor University Medical Center Proceedings, 32*, 5.

16. Hyvarinen, A., & Oja, E. (1997). A fast fixed-point algorithm for independent component analisis. *Neuronal Computation, 9*, 1483–1492.

17. Imai, K., & Kitakoji, S. (2006). Gastric arrhythmia and nausea of motion sickness induced in healthy Japanese subjects viewing an optokinetic rotating drum. *Journal of Physiological Sciences, 56*(5), 341–345.

18. James, C. J., & Hesse, C. W. (2005). Independent component analysis for biomedical signals. *Physiological Measurement, 26*(1), R15.

19. Kara, S., Dirgenali, F., & Okkesim, S. (2006). Detection of gastric dysrhythmia using WT and ANN in diabetic gastroparesis patients. *Computers in Biology and Medicine, 36*(3), 276–290.

20. Koch, K. L., & Robert, M. S. (1994). Electrographic data acquisition and Analysis Electrography: Principles and applications. *Raven Press, 1*, 31–44.

21. Koch, K. L., & Stern, R. M. (1994). Nausea and vomiting and gastric dysrhythmias. In *Electrography: Principles and applications* (Vol. 1, pp. 309–328). New York: Raven Press.

22. Komorowski, D. (2018). EGG DWPack: System for multi-channel electrogastrographic signals recording and analysis. *Journal of Medical Systems, 42*(11), 201.

23. Kovac, A. L. (2018). Updates in the management of postoperative nausea and vomiting. *Adv Anesth, 36*(1), 81–97. https://doi.org/10.1016/j.aan.2018.07.004.

24. Kreis, M. E. (2006). Postoperative nausea and vomiting. *Autonomic Neuroscience: Basic and Clinical, 129*, 86–91.

25. Lee, Y. Y., Kim, K. H., & Yom, Y. H. (2007). Predictive models for post operative nausea and vomiting in patients using patient controlled analgesia. *The Journal of International Medical Research, 35*, 497–507.

26. Liang, H., & Lin, Z. (2001). Detection of delayed gastric emptying from electrogastrograms with support vector machine. *IEEE Transactions on Biomedical Engineering, 48*(5), 601–604.

27. Lior, R., & Maimon, O. (2005). Clustering methods. In *Data mining and knowledge discovery handbook* (pp. 321–352). New York: Springer.

28. Omer, E., McElmurray, L., Kedar, A., Hughes, M. G., Cacchione, R., & Abell, T. L. (2014). Proximal gastric measurements may predict clinical response in patients with nausea and vomiting undergoing temporary endoscopic gastric electrical stimulation. *Gastroenterology, 146*(5, Supplement 1), S-267.

29. Peng, C., Qian, X., & Ye, D. (2007a). Electrogastrogram extraction using independent component analysis with references. *Neural Computing and Applications, 16*, 581–587.

30. Peng, S. Y., Wu, K. C., Wang, J. J., Chuang, J. H., Peng, S. K., & Lai, Y. H. (2007b). Predicting postoperative nausea and vomiting with the application of an artificial neural network. *British Journal of Anaesthesia, 98*(1), 60–65.

31. Prokešová, J., & Dolina, J. (2009). Clinical application of electrogastrography. *Scripta Medica Facultatis Medicae Universitatis Brunensis Masarykianae, 82*(4), 235–238.

32. Sinclair, D. R., Chung, F., & Mezei, G. (1999). Can postoperative nausea and vomiting be predicted? *Anesthesiology, 91*, 109–118.

33. Smout, A. J. P. M., Jebbink, H. J. A., & Samsom, M. (1994). Acquisition and analysis of electrogastrographic data. In *Electrography: Principles and applications* (Vol. 1, pp. 3–28). New York: Raven Press.

34. Somarajan, S., Muszynski, N. D., Russell, A., Gorman, B., Acra, S., Cheng, L. K., & Bradshaw, L. A. (2016). Sa1717 high-density electrogastrogram identifies spatial dysrhythmias in adolescent patients with chronic idiopathic nausea: A preliminary study. *Gastroenterology, 150*(4, Supplement 1), S356.

35. Stern, R. M., Koch, K. L., & Andrews, P. (2011). *Nausea: Mechanisms and management.* New York: Oxford University Press.

36. Vapnik, V. (1995). *The nature of statistical learning theory* (p. 188). New York: Springer.
37. Yin, J., & Chen, J. D. Z. (2013). Electrogastrography: Methodology, validation and applications. *Journal of Neurogastroenterology and Motility, 19*(1), 5–17.
38. Yu, S. N., & Chou, K. T. (2009). Selection of significant independent components for ECG beat classification. *Journal Expert Systems with Applications, 36*(2), 2088–2096.

Random Forest Algorithm for Prediction of HIV Drug Resistance

Letícia M. Raposo, Paulo Tadeu C. Rosa, and Flavio F. Nobre

Abstract Random forest algorithm is a popular choice for genomic data analysis and bioinformatics research. The fundamental idea behind this technique is to combine many decision trees into a single model and use the random subspace method for selection of predictor variables. It is a nonparametric algorithm, efficient for both regression and classification problems, and has a good predictive performance for many types of data. This chapter describes the general characteristics of the random forest algorithm, showing, in practice, a comprehensive application of how this approach can be applied to predict HIV-1 drug resistance. The random forest results were compared to the other two models, logistic regression and classification tree, and presented lower variability in its results, showing to be a classifier with greater stability.

Keywords Classification · HIV drug resistance · Random forest

1 Introduction

Machine learning techniques have been commonly applied to provide solutions to biological questions. Algorithms such as artificial neural networks, support vector machines, and random forest (RF) have played an important role in different tasks including genetic studies [1, 2], proteomics [3, 4], drug development [5, 6], cancer classification [7], clinical decision-making [8], and biomarker identification [3, 9].

In particular, the random forest (RF) algorithm [10] is a great choice in the context of genomics and bioinformatics research. The original algorithm is an ensemble method based on the combination of numerous decision trees using the random subspace method proposed by Tin Ho [11] and bagging [12]. RF is

L. M. Raposo (✉) · P. T. C. Rosa · F. F. Nobre
Programa de Engenharia Biomédica, Universidade Federal do Rio de Janeiro, Rio de Janeiro, Brazil
e-mail: raposo@peb.ufrj.br

© Springer Nature Switzerland AG 2020
M. R. Ortiz-Posadas (ed.), *Pattern Recognition Techniques Applied to Biomedical Problems*, STEAM-H: Science, Technology, Engineering, Agriculture, Mathematics & Health, https://doi.org/10.1007/978-3-030-38021-2_6

nonparametric, efficient for both regression and classification problems, and has a good predictive performance for many types of data. It works well with high-dimensional feature space (the so-called "small n large p" problem), it can be applied in difficult scenarios with highly correlated predictors, and it is able to identify the most important variables in the model [13, 14]. These unique advantages have made RF a popular choice in data mining.

RF has been successfully applied in gene expression [13, 15, 16], proteomics [17], and metabolomics [18–20] research. Additionally, some studies have used RF to predict a phenotype output from genotype data [21–23]. The idea is to apply a technique that learns from the data a pattern for each phenotype. The resulting model is then used to determine the phenotype for a new example. For instance, RF has been shown to be very effective in predicting the drug resistance nature of human immunodeficiency virus (HIV) variants based on their genotypes [24–29].

HIV drug resistance is one of the major obstacles to a successful HIV therapy and is largely responsible for the progressive reduction in the potency of the components of the therapeutic regimen. The use of laboratory tests to detect this problem has played an important role in the decision-making regarding the therapies to be adopted by HIV+ individuals. The most widely used approaches in detecting resistance to antiretrovirals (ARV) are phenotyping and genotyping. Phenotypic assays provide a direct measure of drug susceptibility; however, these tests are expensive, labor-intensive, and time-consuming. Genotypic tests determine the target gene sequence of the drug and use this information to infer ARV susceptibility [30]. This method is more commonly used in clinical practice, since it is generally less costly, is rapidly processable, and is less laborious than phenotyping [31–33].

This chapter aims to describe the general characteristics of RF and how it can be used for the prediction of HIV drug resistance. Section 2 gives a background on the RF algorithm focusing on sample generation, tree construction, randomization, and decision-making. Section 3 shows the error estimation. Section 4 discusses the two most popular variable importance measures. The next section discusses proximities and their utilities followed by a comprehensive application of how RF can be applied to HIV subtype 1 (HIV-1) drug resistance in Sect. 6. Finally, Sect. 7 provides some concluding views.

2 Algorithmic Framework for Random Forest

RF is an ensemble of tree-based learners, constructed originally using the classification and regression tree (CART) algorithm [34] with introduced randomization. Decision trees are a hierarchical structure composed of internal and terminal nodes connected by branches. The tree construction starts with the root node, comprising all the data from the training set. The CART algorithm recursively searches among all predictors and possible split points whose will provide the best binary split of data in such a way as to minimize a measure of node impurity (splitting criterion). To each split, two descending nodes are formed: the left, which receives the data

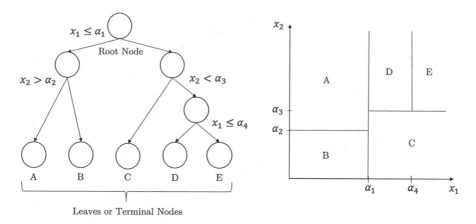

Fig. 1 On the right, an example of a decision tree constructed with two predictor variables. After reaching a predefined stopping criterion, the tree presented five terminal nodes. On the right is illustrated the five decision regions

that satisfies the logical test of the internal node, and the right, which receives the data that does not satisfy the test. These nodes are the decision-making units that evaluate through binary tests that will be the next descending nodes. After reaching a stopping criterion in the division process (e.g., a minimum number of training instances assigned to each terminal node), the final partition splits the predictor space into hyper-rectangles called leaves or terminal nodes, associated with a label or value, as shown in Fig. 1. Given a new observation, the tree is traversed along the tree from the root to the leaf, passing through the decision nodes.

The construction of the original RF technique suggested by Breiman [10] can be described in the following steps:

1. Draw B independent bootstrap samples of size n from the training data L.
2. Fit a tree T_b for each bootstrapped data. At each internal node, select m predictors at random from the p available predictors for splitting. Each tree is grown without pruning.
3. Aggregate information from the B trees for new data prediction by unweighted voting for classification and unweighted averaging for regression.

The general functioning of the original RF algorithm is illustrated in Fig. 2.

2.1 Bootstrap Samples

The RF builds multiple CART trees from a training set using a concept called bootstrapping aggregation (bagging). This meta-algorithm is a special case of the model averaging approach proposed by Breiman [12] that consists of developing

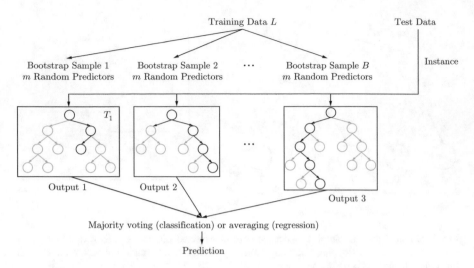

Fig. 2 The general structure of the original random forest algorithm

several models from bootstrap samples. This technique reduces the variance of the final model and helps avoid overfitting, working as a solution to the instability observed in CART trees [22, 35].

Each independent bootstrap sample is produced by random sampling with replacement from the training data, resulting in several samples of the same size as the original training set. This means that a single observation has a chance to appear multiple times within the same sample. In general, about two-thirds of the observations will be sampled (with replicates) and one-third will be left out. Given a training data L with n observations, since all bootstrap samples have the same size, the probability of an observation not being chosen is $(1 - 1/n)$. As this is repeated n times, we have a probability equal to $(1 - 1/n)^n$. With n tending to a large enough number, this value converges to e^{-1}, which is equivalent to approximately 0.368. Thus, approximately 36.8% of the observations will not be considered in the bootstrap samples. The bootstrap technique has several interesting statistical properties, and for its better contextualization and details, see [36].

Each bootstrap sample is used as the training set to construct a decision tree. Observations not included in each bootstrap set, called out-of-bag (OOB), are very useful for estimating the prediction error of the model associated with each bootstrap sample (OOB estimation); see Sect. 3 for details.

2.2 Construction of Trees and Sampling of Variables

After drawing the B bootstrap samples, each tree is constructed from splitting rules that aim to maximize impurity reduction. When not performing any pruning step, the original RF algorithm will obtain low-bias trees but with high variance. This variance is reduced by bagging.

At each step of constructing a single decision node, the RF algorithm selects a random subset of $m \ll p$ predictor variables and searches for the optimal split over these latter variables only. The number of variables is fixed, and it is usually recommended $m = \lfloor \sqrt{p} \rfloor$ for classification problems, with minimum node size equal to one and $m = \lfloor p/3 \rfloor$ for regression, with a minimum node size equal to five [10].

At each split s of a node t into its two descendants (t_L and t_R), each of the m randomly selected predictor variables is evaluated according to a splitting criterion, as seen previously for the CART algorithm. The idea is to look for, among the m variables, one that will maximize the decrease

$$\Delta i\,(s,t) = i(t) - pr_L i\,(t_L) - pr_R i\,(t_R) \tag{1}$$

of some impurity criterion $i(t)$, and $pr_L = n_{t_L}/n$ and $pr_R = n_{t_R}/n$.

For classification trees, a common impurity criterion is the so-called Gini index. This index measures how often a randomly selected observation from the set would be incorrectly classified if it was randomly labeled according to the distribution of class labels from the data set. The Gini index is defined as follows:

$$G(t) = 1 - \sum_{c=1}^{C} pr_c \tag{2}$$

where C is the number of classes, and pr_c is the proportion of observations belonging to class c in the node.

When a variable is examined, the weighted mean impurities of implied descendent nodes are subtracted from $G(t)$, and the variable that results in the greatest impurities decrease is selected. The Gini index is minimum ($G(t) = 0$) if the subset comprises a single class and maximum when C classes are equally distributed.

For regression trees, if the response values at the node are y_i, a classic impurity measure is mean squared error (MSE):

$$\mathrm{MSE} = \frac{1}{n} \sum_{i=1}^{n} (y_i - \overline{y})^2 \tag{3}$$

where $\overline{y} = \frac{1}{n}\sum_{i=1}^{n} y_i$ is the mean response for the training observations at the node.

The predictor that produces the smallest sum of the splitting criterion values weighted by the proportion of the number of observations in each node will be chosen for the division. The procedure continues recursively until a stopping criterion is reached [34, 37].

By randomly sampling the variables and the input data, each tree is effectively an independent model, not being influenced by other decision trees. Thus, the resulting model tends not to overfit the training data set, providing considerable robustness to noise and outliers.

Besides the advantage of introducing randomness in the model, limiting the number of variables evaluated for a split, the randomization avoids the "small n large p" problem, reduces the correlation between trees, and also offers a computational efficiency.

2.3 Decision

After building B decision trees, each one has equal weight in the final decision-making process. In classification problems, the final label is the majority vote of the trees, and the prediction at a new point x is given by the following:

$$\hat{f}(x) = \operatorname*{argmax}_{y} \sum_{b=1}^{B} I\left(y = \hat{T}_b(x)\right) \tag{4}$$

For regression problems, the result is the average value on the set of trees, given by the following:

$$\hat{f}(x) = \frac{1}{B} \sum_{b=1}^{B} \hat{T}_b(x) \tag{5}$$

where $\hat{T}_b$ is the prediction of the response variable at x using the bth tree.

3 Error Estimation

For each tree, the set of observations which is not used to construct a tree is denoted as OOB observations. These samples are particularly useful for estimating generalization errors, acting as an internal validation data set in the choice of tuning parameters and in the evaluation of the variable importance.

Let L_b represent the bth bootstrap sample and $\hat{T}_b(x)$ denote the prediction at x from the bth tree. Also, let $\mathcal{B}_i = \{b : (x_i, y_i) \notin L_b\}$ and let B_i be the cardinality of $\mathcal{B}_i$.

For classification with a zero-one loss, the generalization error rate is assessed using the OOB error rate:

$$E_{\text{OOB}} = \frac{1}{n} \sum_{i=1}^{n} I\left(y_i \neq \hat{f}_{\text{OOB}}(x_i)\right) \tag{6}$$

and $\hat{f}_{\text{OOB}}(x_i)$ is given by the following:

$$\hat{f}_{\text{OOB}}(x_i) = \underset{y}{\text{argmax}} \sum_{b \in \mathcal{B}_i} I\left(\hat{T}_b(x_i) = y_i\right) \tag{7}$$

For regression with a squared error loss, the generalization error is typically estimated by OOB MSE:

$$\text{MSE}_{\text{OOB}} = \frac{1}{n} \sum_{i=1}^{n} \left(y_i - \hat{f}_{\text{OOB}}(x_i)\right)^2 \tag{8}$$

where $\hat{f}_{\text{OOB}}(x_i)$ is the OOB prediction for observation i given by the following equation:

$$\hat{f}_{\text{OOB}}(x_i) = \frac{1}{B_i} \sum_{b \in \mathcal{B}_i} \hat{T}_b(x_i) \tag{9}$$

Tibshirani [38], Wolpert, and Macready [39] pointed out that the error calculated with the OOB observations is an unbiased estimator of the prediction error, often producing statistics that are as stable as or even more accurate than k-fold cross-validation estimates, also including the advantage of being computationally efficient. Breiman [40] showed that the OOB estimation is as good as that obtained with a test set of the same size as the training set.

In addition to working as an internal validation step, the OOB error can be applied in the choice of the tuning parameters, searching for the ones which minimize this error. The tuning parameters are the number of trees, B (also called *ntree*), and the size of the variable subset, m (also called *mtry*). Breiman [10] showed that as the number of trees increases, the generalization error for RF converges almost to the limit. Thus, the concern is to define a number of trees that are not too small. For this, it is suggested that several increasing values be evaluated until some measure of interest, e.g., the OOB error, stabilizes.

It is also possible to find the best value of m by using the OOB error. If the chosen value of m is too small, it is possible that none of the randomly selected predictor variables is relevant, producing inadequate splits and consequently trees with poor predictive ability. On the other hand, if the subset contains a large number of input variables, it is likely that the same strong predictors will be repeatedly selected by the splitting criterion, and consequently, the variables with smaller effects will have almost no chance of being chosen [41]. Díaz-Uriarte and Alvarez de Andrés [13] showed in an empirical study the robustness of RF to variations in its parameters.

4   Variable Importance Measures

Although the interpretability of an RF is not as straightforward as that of a single decision tree, in which predictor variables closer to the root node are more influential, RF provides alternative measures to evaluate the variable importance.

Through variable importance measures, it is possible to identify which predictor variables are the most important to make predictions and, from this information, reduce the large set of variables to those that contain more information for the problem. This approach has made it possible to use RF not only as predictors but also as a strategy to identify relevant features or perform variable selection.

Several techniques for computing variable importance scores have already been presented in the literature [42–44]. Among them, the simplest is to count the number of times the variable appears in the set of decision trees. Those with the highest numbers of appearances are considered the most important for the problem.

Two other very popular strategies are *impurity importance* and *permutation importance*. These approaches, proposed by [10, 45], consider measures of importance in decision-making.

4.1   *Impurity Importance*

A common alternative to quantify variable importance is to evaluate the average decrease in the impurity. Variables used in divisions that provide a large decrease in impurity are considered important. When a tree T is constructed, at each node t, the decrease in impurity measure is calculated for variable x used to make the split. The variable importance is then given by the average decrease in impurity measure in the forest, where the variable x is used to split a node:

$$\text{Imp}(x) = \frac{1}{n_T} \sum_{T} \sum_{t \in T : v(s_t) = x} pr(t) \Delta i \, (s_t, t) \qquad (10)$$

where $pr(t) = n_t/n$ and represents the proportion of samples reaching nodes t, and $v(s_t)$ is the variable used in split s_t. This strategy is also called *mean decrease impurity* importance.

For classification, the impurity is frequently measured by the Gini index (*Gini importance* or *mean decrease Gini*). For regression, the same procedure with MSE as impurity measure is commonly used.

4.2 Permutation Importance

Another measure of variable importance is based on a permutation test. The main idea is that, if the variable is not important for the problem, rearranging the values of that variable will not interfere with the prediction accuracy. First, the OOB sample is passed down the tree, the predicted values are computed, and the prediction error is registered. Following, the values of the variable x in the OOB sample are randomly permutated, preserving all other variables. These modified OOB data are passed down the tree, and the predicted values are also computed. The new prediction error in the permutated data is also measured. The average of the difference between the OOB error with and without permutation over all the trees is the permutation importance of the variable x. This procedure is repeated for all variables of interest. The larger the value of permutation importance of a variable, the more important the variable is for the overall prediction accuracy. For classification, this variable importance measure is also called *mean decrease accuracy* or *mean increase error*. For regression problems, the same approach is used but with the MSE instead of error rates.

Studies have shown that the impurity importance can be biased in favor of variables with many possible split points, i.e., categorical predictors with many categories and continuous predictors [14, 34], and in favor of variables with high category frequencies [46]. The permutation importance is more robust to these situations and thus is commonly preferred [47–49]. However, permutation importance is very computationally intensive for high-dimensional data. To overcome these limitations, strategies have been presented as a way to avoid this possible distortion; see [5] for details.

5 Proximities

From the trained trees, the RF is able to provide information about pairwise proximity between observations of the data set. For each tree built, the data, both training, and OOB, are run down the tree. If two observations x_i and x_j occupy the same terminal node of the tree, their proximity is increased by 1. If they are never in the same terminal node, their proximity will be 0. These values are summed and divided by the number of trees in RF [42]. All samples' proximities create a matrix, with size $n \times n$. Proximity matrix is symmetric with 1 on the main diagonal and in 0 to 1 off the diagonal. The greater the proximity value is, the more similar samples are.

As proposed by Breiman [10], the sample proximity from RF could be used to remove outliers. The average proximity from case x_i in class c to the rest of the training data class $c(k)$ is defined as follows:

$$\overline{P}(x_i) = \sum_{class(k)=c} prox^2(x_i, k) \tag{11}$$

For each observation, the average proximity to the other samples in the class is calculated. The raw outlier measure for case x_i is the sample size (n) divided by this average proximity $\overline{P}(x_i)$. Within each class, the median of the raw measures and their absolute deviation from the median are calculated. The median is subtracted from each raw measure, and this result is divided by the absolute deviation to get the final outlier measure. If this value is large, the proximity is small, and the case is determined as an outlier [50, 51].

Proximities are also used for missing value imputation and producing low-dimensional views of the data. To impute missing values, an initial forest is grown replacing any missing value with the median (noncategorical variable) or the mode (categorical variable) of that predictor for all other cases. Proximities are computed and new imputations are obtained. For continuous predictors, the imputed value is the weighted average of the non-missing observations; the weights are given by the proximities. For categorical predictors, the imputed value is the category with the largest average proximity. Generally, this procedure is repeated four to six times to ensure stable imputations [37, 51].

Through proximity matrices, RF provides a way of visualizing data, which is often not possible due to high dimensionality. An RF proximity distance matrix with multidimensional scaling (MDS) is obtained to get two-dimensional or three-dimensional plots. Each point in the graph symbolizes one of the observations, and the distances between the points reproduce as far as possible the proximity-based distances. Some of the applications of this plot are in the selection of subgroups of cases (clusters) that almost always remain together in trees or also in the identification of outliers that are almost always alone in a terminal node [37].

6 Application on HIV-1 Drug Resistance Problem

6.1 Background

In this section, we considered the HIV-1 resistance problem as an example of application of the RF algorithm. We developed a RF classifier to predict HIV-1 resistance to Lopinavir (LPV), an inhibitor of HIV protease. All the analysis were performed in R software. Scripts were available in https://raw.githubusercontent.com/leticiaraposo/hivdb/master/RScript to make the example reproducible.

The data were obtained from the Stanford HIV Drug Resistance Database (HIVDB) (https://hivdb.stanford.edu/pages/genopheno.dataset.html) in February 2019. The data have been deposited and are found in this link. This public available source is based on genotype-phenotype pairs and has been used in many HIV resistance studies [52–54]. HIVDB shows a set of sequences of enzymatic targets of antiretroviral therapy with information of amino acid mutations associated with specific positions and a phenotyping value of drug susceptibility. For training and testing, all HIV-1 subtype B observations containing results of LPV susceptibility test analyzed using the PhenoSense® (a specific phenotyping assay) were selected.

The data were preprocessed. The predictor variables were represented by the 99 positions of the HIV protease enzyme. Initially, each position is represented by an amino acid, which is equal to that of the reference sequence or different when it is a mutation. Mutations were defined when there was a difference between the amino acids of the consensus subtype B sequence and the amino acids found in the sequencing. Observations whose sequences showed deletions, insertions, stop codons, mixtures of amino acids, and unknown amino acids were excluded from the data set. Duplicated sequences were included only once. These criteria led to 591 amino acid sequences of the protease enzyme of the pol gene of HIV-1 subtype B from infected patients.

Based on the information of these databases, both regression and classification models can be constructed to predict HIV-1 resistance to ARVs. In regression models, the genotype-phenotype pairs can be used directly as training data, and the purpose of the model is to predict the phenotype value for a specific drug from the mutational patterns. When considering classification problems, the response variable (phenotype value) needs to be converted into class labels before being utilized. The method commonly used for the interpretation of phenotypic data is based on biological cutoffs. There is a specific cutoff for each drug that determines when the virus is considered as "drug-susceptible" or as "drug-resistant" [55, 56]. Most of the time, the problem is binary (resistant or susceptible), but it can also be divided into three classes with the addition of the "intermediate resistance" category.

In this example, we have considered a classification problem and transformed the response variable originally expressed as the fold change defined as the ratio of the half maximal inhibitory concentration (IC50) of a mutant and a wild-type standard assayed by the PhenoSense assay into a categorical variable with two classes. IC50 represents the concentration of ARV inhibiting 50% of viral replication compared to cell culture experiments without ARV. The cutoff value for LPV is 9.0 as indicated by [57]. In this way, if the fold change is less than the cutoff value, the mutant was codified as nonresistant or susceptible to the LPV. Otherwise, values greater than or equal to 9.0 indicated resistance to LPV. Of the 591 observations, 366 were nonresistant to LPV and 225 resistant.

For the application of a regression or classification algorithm, it is necessary that the genotype represented by the amino acids is properly encoded. For this, several strategies have been applied in the literature. One of the simplest consists of representing each amino acid by a 20-bit vector with 19 bits set to zero and one bit set to one [58, 59]. Another alternative is to convert each amino acid to positive integers [60] or only assign 0 when the amino acid is the same as the reference sequence and 1 when the mutation occurred [61]. Many studies have used information on the physicochemical and biochemical properties of amino acids [58, 62, 63]. Over 500 indexes have already been reported in the AA index database [64].

In this application, the Kyte-Doolittle hydrophobicity scale was chosen to numerically represent the amino acids of each predictor variable. This scale has been used in other studies [65, 66]. Hydrophobicity scales are the measures of hydrophobicity for amino acid residues. The more positive the valueis, the more hydrophobic the residue. The mutation affects the hydrophobicity scale, thereby disrupting the overall structure. After the encoding, the constant and almost constant predictors across samples (called zero and near-zero variance predictors) were excluded using the *nearZeroVar* function from R-package *caret* [67], resulting in 30 input variables.

The set of 591 available amino acid sequences was divided into a training set of 413 sequences and a test set of 178 sequences. In the training group, 256 patients showed no resistance to LPV, whereas 157 were resistant. In the test group, 110 patients were resistant and 68 showed no resistance. The training set was used to tune the parameters, construct the trees, and evaluate the generalization of RF. The test set was only used to assess the performance of the final model.

We trained RFs for the prediction of LPV resistance using the R-package *randomForest* [43]. We used the default values for the number of features (*mtry*) and the number of trees (*ntree*). For this configuration (*ntree* = 500, *mtry* = 4), the OOB estimate of error rate was 8.23%.

We compared RF to two other algorithms commonly used: classification trees (CT) and logistic regression (LR). No "tuning" of the classification procedures was carried out. The importance of each predictor variable for RF, i.e., sequence position, for the correct classification was assessed by the mean decrease accuracy, as explained in Sect. 4.2 (Fig. 3).

For CT, the variables split on in the highest nodes were considered to be the most important for that procedure. For LR, backward elimination was carried out using the Akaike information criterion (AIC), and the retained variables were ranked by the p-value.

The top 10 positions ranked for RF were in the following order: 54, 46, 82, 33, 10, 71, 84, 76, 47, and 32, respectively. For CT, the positions were in this order, 46, 54, 71, 82, 10, 84, 76, 47, 33, and 77. For the LR, according to the p-value, the positions were as follows: 46, 36, 82, 90, 33, 47, 71, 10, 76, and 84.

Eight positions (10, 33, 46, 47, 71, 76, 82, 84) were ranked among the top 10 positions by the three techniques. When RF was compared with CT, nine positions were shared. Six positions appeared exclusively for each technique when only the first 10 variables were evaluated. They were 32 for RF, 77 for CT, and 36 and 90 for LR.

As reported by HIVDB [68], the positions associated with major LPV resistance mutations were 32, 33, 46, 47, 48, 50, 54, 76, 82, 84, and 90. Of these, six of the eight positions shared by the three techniques were well ranked. Positions 10 and 71, although not associated with major protease inhibitor (PI) resistance mutations, are also important for the resistance problem, since they may exhibit accessory mutations associated with reduced susceptibility to PI. Only position 77, selected by CT, is not among the most important positions according to HIVDB. It can be

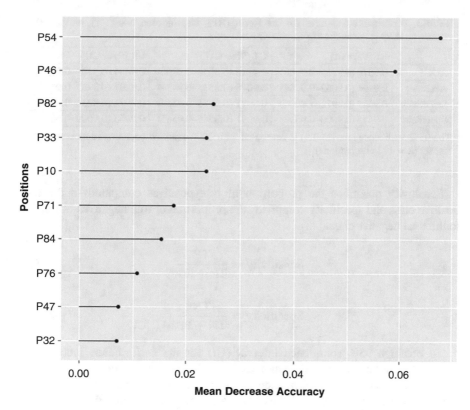

Fig. 3 Variable importance plot for top 10 predictor variables from random forest classifications used for predicting HIV-1 resistance to Lopinavir. The mean decrease in accuracy was used to access the variable importance

observed that the variables RF identified as most important to predict LPV resistance coincided with the mutations reported in the published literature.

The performance of the models was evaluated using accuracy, sensitivity, and specificity (see formulas below). The models were applied to the test set with 178 samples, which was not used at any other stage of the analysis. The area under the ROC curve (AUC) was also obtained. It represents the expected performance as a single scalar.

Accuracy is defined as the proportion of correct classification by the model over the total sample. This index is given by the following formula:

$$\text{Accuracy} = \frac{TP + TN}{TP + FP + TN + FN} \tag{12}$$

where TP, FP, TN, and FN are true positives, false positives, true negatives, and false negatives, respectively.

Table 1 Performance measures for the forest (RF), classification tree (CT), and logistic regression (LR) for prediction of Lopinavir resistance

	RF (95% CI)	CT (95% CI)	LR (95% CI)
AUC	0.9878 (0.9767–0.9988)	0.9386 (0.9059–0.9800)	0.9664 (0.9423–0.9906)
Accuracy	0.9494 (0.9062–0.9766)	0.9045 (0.8515–0.9434)	0.9207 (0.8783–0.9322)
Sensitivity	0.9559 (0.8971–1.0000)	0.8971 (0.8235–0.9559)	0.8971 (0.8235–0.9706)
Specificity	0.9455 (0.9000–0.9818)	0.9091 (0.8545–0.9545)	0.9455 (0.9000–0.9818)

The 95% confidence interval (CI) is computed with 2000 stratified bootstrap replicates of the test set ($n = 172$ observations)

Sensitivity quantifies the proportion of true positives compared to the total positive class and specificity comprises the proportion of true negatives in relation to the total negative class.

$$\text{Sensitivity} = \frac{TP}{TP + FN} \tag{13}$$

$$\text{Specificity} = \frac{TN}{TN + FP} \tag{14}$$

To obtain a 95% confidence interval (CI) for the performance index, 2000 bootstrap samplings with stratification of the test set were performed.

When evaluating the performance of the models, in general, the values were close between the techniques (Table 1). However, although all 95% CI showed an intersection between their values, the RF presented all median performance measures superior to the alternative classifiers.

Figure 4 shows that when performing 2000 stratified bootstrap replicates, the three classifiers display intersection of their AUC values. However, RF presented the lowest variability in its results (95% CI, 0.9741–0.9982) when compared to CT and LR models. The original motivation for the development of RF was to increase the classification accuracy and stability of classification trees. This can be observed in this example.

Overall, the three techniques explored in this example showed good performance in the prediction of HIV-1 resistance to LPV. The RF presented lower variability in its results, showing to be a classifier with greater stability.

7 Conclusion

In summary, RF works as a combination of decision trees, being an effective predictive tool for antiretroviral resistant identification. In this chapter, we show how the tool can be applied in the prediction of HIV-1 subtype B resistance to the drug Lopinavir.

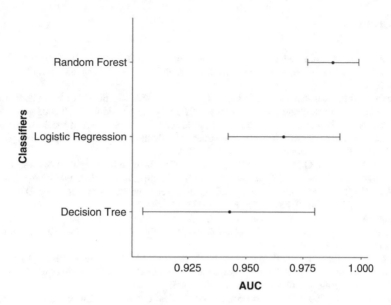

Fig. 4 Variability of AUC values obtained from 2000 bootstrap samples for the random forest, classification tree, and logistic regression techniques

RF has the following advantages: (1) It can be used to construct models for both classification and regression problems. (2) Its default hyperparameters often produce a good prediction result. (3) The relative importance it assigns to the input features is easy to view. (4) OOB data can be used to estimate the RF generalization error and can also estimate the importance of individual variables. (5) The combination of bagging and random selection of features to split allows RF to better tolerate perturbations of the data. (6) RF can manipulate continuous variables and categorical variables. (7) RF does not make any distributive assumptions about predictor or response variables and can deal with situations in which the number of predictor variables far exceeds the number of observations. (8) RF produces automatic proximities – measures of similarity among data points – which can be used to impute missing values.

In general, RF has proven to be a simple and flexible tool, although it has its limitations. It is more complex to visualize the model or to obtain a direct understanding of the predicted outputs, i.e., the values produced by RF do not have simple representations such as formula (e.g., logistic regression) or graphical scheme (e.g., classification trees) that characterize the classification function. This lack of simple representation can make interpretation difficult.

Despite the limitations, RF has been used successfully for a wide variety of applications, including some in the biomedical field. Many improvements can be applied to this technique as a way to overcome some drawbacks. This suggests that RF will still be widely used successfully in different scenarios.

References

1. Mutalib, S., & Mohamed, A. (2011). A brief survey on GWAS and ML algorithms. In *2011 11th International Conference on Hybrid Intelligent Systems (HIS)* (pp. 658–661). Piscataway: IEEE.
2. Szymczak, S., Biernacka, J. M., Cordell, H. J., et al. (2009). Machine learning in genome-wide association studies. *Genetic Epidemiology, 33*, S51–S57. https://doi.org/10.1002/gepi.20473.
3. Swan, A. L., Mobasheri, A., Allaway, D., et al. (2013). Application of machine learning to proteomics data: Classification and biomarker identification in postgenomics biology. *OMICS, 17*, 595–610. https://doi.org/10.1089/omi.2013.0017.
4. Barla, A., Jurman, G., Riccadonna, S., et al. (2007). Machine learning methods for predictive proteomics. *Briefings in Bioinformatics, 9*, 119–128. https://doi.org/10.1093/bib/bbn008.
5. Wale, N. (2011). Machine learning in drug discovery and development. *Drug Development Research, 72*, 112–119. https://doi.org/10.1002/ddr.20407.
6. Lima, A. N., Philot, E. A., Trossini, G. H. G., et al. (2016). Use of machine learning approaches for novel drug discovery. *Expert Opinion on Drug Discovery, 11*, 225–239. https://doi.org/10.1517/17460441.2016.1146250.
7. Kourou, K., Exarchos, T. P., Exarchos, K. P., et al. (2015). Machine learning applications in cancer prognosis and prediction. *Computational and Structural Biotechnology Journal, 13*, 8–17. https://doi.org/10.1016/J.CSBJ.2014.11.005.
8. Kononenko, I. (2001). Machine learning for medical diagnosis: History, state of the art and perspective. *Artificial Intelligence in Medicine, 23*, 89–109. https://doi.org/10.1016/S0933-3657(01)00077-X.
9. Najami, M., Abedallah, N., & Khalifa, L. (2014). Computational approaches for bio-marker discovery. *Journal of Intelligent Learning Systems and Applications, 6*, 153–161. https://doi.org/10.4236/jilsa.2014.64012.
10. Breiman, L. (2001). Random forests. *Machine Learning, 45*, 5–32. https://doi.org/10.1023/A:1010933404324.
11. Ho, T. K. (1998). The random subspace method for constructing decision forests. *IEEE Transactions on Pattern Analysis and Machine Intelligence, 20*, 832–844. https://doi.org/10.1109/34.709601.
12. Breiman, L. (1996). Bagging predictors. *Machine Learning, 24*, 123–140. https://doi.org/10.1007/BF00058655.
13. Díaz-Uriarte, R., & Alvarez de Andrés, S. (2006). Gene selection and classification of microarray data using random forest. *BMC Bioinformatics, 7*, 3. https://doi.org/10.1186/1471-2105-7-3.
14. Strobl, C., Boulesteix, A.-L., Zeileis, A., & Hothorn, T. (2007). Bias in random forest variable importance measures: Illustrations, sources and a solution. *BMC Bioinformatics, 8*, 25. https://doi.org/10.1186/1471-2105-8-25.
15. Hsueh, H.-M., Zhou, D.-W., & Tsai, C.-A. (2013). Random forests-based differential analysis of gene sets for gene expression data. *Gene, 518*, 179–186. https://doi.org/10.1016/J.GENE.2012.11.034.
16. Wu, X., Wu, Z., & Li, K. (2008). Identification of differential gene expression for microarray data using recursive random forest. *Chinese Medical Journal, 121*, 2492–2496.
17. Montaño-Gutierrez, L. F., Ohta, S., Kustatscher, G., et al. (2017). Nano Random Forests to mine protein complexes and their relationships in quantitative proteomics data. *Molecular Biology of the Cell, 28*, 673–680. https://doi.org/10.1091/mbc.e16-06-0370.
18. Cao, Z. W., Han, L. Y., Zheng, C. J., et al. (2005). Computer prediction of drug resistance mutations in proteins. *Drug Discovery Today, 10*, 521–529. https://doi.org/10.1016/S1359-6446(05)03377-5.
19. Chen, T., Cao, Y., Zhang, Y., et al. (2013). Random forest in clinical metabolomics for phenotypic discrimination and biomarker selection. *Evidence-based Complementary and Alternative Medicine, 2013*, 298183. https://doi.org/10.1155/2013/298183.

20. Abdullah, M. N., Yap, B. W., Zakaria, Y., & Abdul Majeed, A. B. (2016). Metabolites selection and classification of metabolomics data on Alzheimer's disease using random forest. In M. Berry, A. Hj Mohamed, & B. Yap (Eds.), *Soft computing in data science. SCDS 2016. Communications in Computer and Information Science* (Vol. 652, pp. 100–112). Singapore: Springer.
21. Goldstein, B. A., Hubbard, A. E., Cutler, A., & Barcellos, L. F. (2010). An application of Random Forests to a genome-wide association dataset: Methodological considerations & new findings. *BMC Genetics, 11*, 49. https://doi.org/10.1186/1471-2156-11-49.
22. Goldstein, B. A., Polley, E. C., & Briggs, F. B. S. (2011). Random forests for genetic association studies. *Statistical Applications in Genetics and Molecular Biology, 10*, 32. https://doi.org/10.2202/1544-6115.1691.
23. Nguyen, T.-T., Huang, J., Wu, Q., et al. (2015). Genome-wide association data classification and SNPs selection using two-stage quality-based Random Forests. *BMC Genomics, 16*, S5. https://doi.org/10.1186/1471-2164-16-S2-S5.
24. Shen, C., Yu, X., Harrison, R. W., & Weber, I. T. (2016). Automated prediction of HIV drug resistance from genotype data. *BMC Bioinformatics, 17*, 278. https://doi.org/10.1186/s12859-016-1114-6.
25. Heider, D., Verheyen, J., & Hoffmann, D. (2010). Predicting Bevirimat resistance of HIV-1 from genotype. *BMC Bioinformatics, 11*, 37. https://doi.org/10.1186/1471-2105-11-37.
26. Wang, D., Larder, B., Revell, A., et al. (2009). A comparison of three computational modelling methods for the prediction of virological response to combination HIV therapy. *Artificial Intelligence in Medicine, 47*, 63–74. https://doi.org/10.1016/J.ARTMED.2009.05.002.
27. Khalid, Z., & Sezerman, O. U. (2016). Prediction of HIV drug resistance by combining sequence and structural properties. *IEEE/ACM Transactions on Computational Biology and Bioinformatics, 15*, 966–973. https://doi.org/10.1109/TCBB.2016.2638821.
28. Tarasova, O., Biziukova, N., Filimonov, D., et al. (2018). A computational approach for the prediction of HIV resistance based on amino acid and nucleotide descriptors. *Molecules, 23*, 2751. https://doi.org/10.3390/molecules23112751.
29. Revell, A. D., Wang, D., Perez-Elias, M.-J., et al. (2018). 2018 update to the HIV-TRePS system: The development of new computational models to predict HIV treatment outcomes, with or without a genotype, with enhanced usability for low-income settings. *The Journal of Antimicrobial Chemotherapy, 73*, 2186–2196. https://doi.org/10.1093/jac/dky179.
30. Bronze, M., Steegen, K., Wallis, C. L., et al. (2012). HIV-1 phenotypic reverse transcriptase inhibitor drug resistance test interpretation is not dependent on the subtype of the virus backbone. *PLoS One, 7*, e34708. https://doi.org/10.1371/journal.pone.0034708.
31. Beerenwinkel, N., Schmidt, B., Walter, H., et al. (2002). Diversity and complexity of HIV-1 drug resistance: A bioinformatics approach to predicting phenotype from genotype. *Proceedings of the National Academy of Sciences of the United States of America, 99*, 8271–8276. https://doi.org/10.1073/pnas.112177799.
32. Vercauteren, J., & Vandamme, A. M. (2006). Algorithms for the interpretation of HIV-1 genotypic drug resistance information. *Antiviral Research, 71*, 335–342. https://doi.org/10.1016/j.antiviral.2006.05.003.
33. Schutten, M. (2006). Resistance assays. In A. M. Geretti (Ed.), *Antiretroviral resistance in clinical practice*. London: Mediscript.
34. Breiman, L., Friedman, J., Olshen, R. A., & Stone, C. J. (1984). *Classification and regression trees*. Belmont: Wadsworth.
35. Hastie, T., Tibshirani, R., & Friedman, J. (2008). *The elements of statistical learning data mining, inference, And prediction*. New York: Springer.
36. Efron, B., & Tibshirani, R. (1994). *An introduction to the bootstrap*. New York: Chapman & Hall.
37. Cutler, A., Cutler, D. R., & Stevens, J. R. (2012). Random forests. In *Ensemble machine learning* (pp. 157–175). Boston: Springer US.
38. Tibshirani, R., & Tibshirani, R. (1996). *Bias, variance and prediction error for classification rules*. Toronto: University of Toronto.

39. Wolpert, D. H., & Macready, W. G. (1999). An efficient method to estimate bagging's generalization error. *Machine Learning, 35*, 41–55. https://doi.org/10.1023/A:1007519102914.
40. Breiman, L. (1996). *Out-of-bag estimation*. Berkeley, CA.
41. Janitza, S., & Hornung, R. (2018). On the overestimation of random forest's out-of-bag error. *PLoS One, 13*, e0201904. https://doi.org/10.1371/journal.pone.0201904.
42. Breiman, L., & Cutler, A. (2004). *RFtools – for predicting and understanding data*. Berkeley University, Berkeley, CA.
43. Liaw, A., & Wiener, M. (2002). Classification and regression by randomForest. *R News, 2*, 18–22.
44. Janitza, S., Celik, E., & Boulesteix, A.-L. (2018). A computationally fast variable importance test for random forests for high-dimensional data. *Advances in Data Analysis and Classification, 12*, 885–915. https://doi.org/10.1007/s11634-016-0276-4.
45. Breiman, L. (2002). *Manual on setting up, using, and understanding random forests v3.1*. Berkeley, CA.
46. Nicodemus, K. K. (2011). Letter to the editor: On the stability and ranking of predictors from random forest variable importance measures. *Briefings in Bioinformatics, 12*, 369–373. https://doi.org/10.1093/bib/bbr016.
47. Nicodemus, K. K., Malley, J. D., Strobl, C., & Ziegler, A. (2010). The behaviour of random forest permutation-based variable importance measures under predictor correlation. *BMC Bioinformatics, 11*, 110. https://doi.org/10.1186/1471-2105-11-110.
48. Szymczak, S., Holzinger, E., Dasgupta, A., et al. (2016). r2VIM: A new variable selection method for random forests in genome-wide association studies. *BioData Mining, 9*, 7. https://doi.org/10.1186/s13040-016-0087-3.
49. Ziegler, A., & König, I. R. (2014). Mining data with random forests: Current options for real-world applications. *Wiley Interdisciplinary Reviews: Data Mining and Knowledge Discovery, 4*, 55–63. https://doi.org/10.1002/widm.1114.
50. Zhang, J., Zulkernine, M., & Haque, A. (2008). Random-forests-based network intrusion detection systems. *IEEE Transactions on Systems, Man, and Cybernetics – Part C: Applications and Reviews, 38*, 649–659. https://doi.org/10.1109/TSMCC.2008.923876.
51. Breiman, L., & Cutler, A. *Random forests – classification description*. https://www.stat.berkeley.edu/~breiman/RandomForests/cc_home.htm#prox. Accessed 19 Dec 2018.
52. Pawar, S. D., Freas, C., Weber, I. T., & Harrison, R. W. (2018). Analysis of drug resistance in HIV protease. *BMC Bioinformatics, 19*, 362. https://doi.org/10.1186/s12859-018-2331-y.
53. Singh, Y. (2017). Machine learning to improve the effectiveness of ANRS in predicting HIV drug resistance. *Healthcare Informatics Research, 23*, 271. https://doi.org/10.4258/hir.2017.23.4.271.
54. Raposo, L. M. L. M., & Nobre, F. F. F. F. (2017). Ensemble classifiers for predicting HIV-1 resistance from three rule-based genotypic resistance interpretation systems. *Journal of Medical Systems, 41*, 155. https://doi.org/10.1007/s10916-017-0802-8.
55. Geretti, A. M., & National Center for Biotechnology Information (U.S.). (2006). *Antiretroviral resistance in clinical practice*. London: Mediscript Ltd.
56. Winters, B., Montaner, J., Harrigan, P. R., et al. (2008). Determination of clinically relevant cutoffs for HIV-1 phenotypic resistance estimates through a combined analysis of clinical trial and cohort data. *JAIDS Journal of Acquired Immune Deficiency Syndromes, 48*, 26–34. https://doi.org/10.1097/QAI.0b013e31816d9bf4.
57. Reeves, J. D., & Parkin, N. T. (2017). Viral phenotypic resistance assays. In *Antimicrobial drug resistance* (pp. 1389–1407). Cham: Springer International Publishing.
58. Bozek, K., Lengauer, T., Sierra, S., et al. (2013). Analysis of physicochemical and structural properties determining HIV-1 coreceptor usage. *PLoS Computational Biology, 9*, e1002977. https://doi.org/10.1371/journal.pcbi.1002977.
59. Rö Gnvaldsson, T., You, L., & Garwicz, D. (2015). State of the art prediction of HIV-1 protease cleavage sites. *Bioinformatics, 31*(8), 1204–1210. https://doi.org/10.1093/bioinformatics/btu810.

60. Sheik Amamuddy, O., Bishop, N. T., & Tastan Bishop, Ö. (2017). Improving fold resistance prediction of HIV-1 against protease and reverse transcriptase inhibitors using artificial neural networks. *BMC Bioinformatics, 18*, 369. https://doi.org/10.1186/s12859-017-1782-x.
61. Van der Borght, K., Verheyen, A., Feyaerts, M., et al. (2013). Quantitative prediction of integrase inhibitor resistance from genotype through consensus linear regression modeling. *Virology Journal, 10*, 8. https://doi.org/10.1186/1743-422X-10-8.
62. Dybowski, J. N., Riemenschneider, M., Hauke, S., et al. (2011). Improved Bevirimat resistance prediction by combination of structural and sequence-based classifiers. *BioData Mining, 4*, 26. https://doi.org/10.1186/1756-0381-4-26.
63. Riemenschneider, M., Hummel, T., & Heider, D. (2016). SHIVA – a web application for drug resistance and tropism testing in HIV. *BMC Bioinformatics, 17*, 314. https://doi.org/10.1186/s12859-016-1179-2.
64. Kawashima, S., & Kanehisa, M. (2000). AAindex: Amino acid index database. *Nucleic Acids Research, 28*, 374–374. https://doi.org/10.1093/nar/28.1.374.
65. Riemenschneider, M., Cashin, K. Y., Budeus, B., et al. (2016). Genotypic prediction of co-receptor tropism of HIV-1 subtypes A and C. *Scientific Reports, 6*, 24883. https://doi.org/10.1038/srep24883.
66. Heider, D., Dybowski, J. N., Wilms, C., & Hoffmann, D. (2014). A simple structure-based model for the prediction of HIV-1 co-receptor tropism. *BioData Mining, 7*, 14. https://doi.org/10.1186/1756-0381-7-14.
67. Kuhn, M. (2016). *Package "caret."* ftp://cran.r-project.org/pub/R/web/packages/caret/caret.pdf. Accessed 20 Feb 2017.
68. Stanford University – HIV Drug Resistance Database. (2016). *PI resistance notes – HIV Drug Resistance Database*. https://hivdb.stanford.edu/dr-summary/resistance-notes/PI/. Accessed 27 Dec 2018.

Analysis of Cardiac Contraction Patterns

Luis Jiménez-Ángeles, Verónica Medina-Bañuelos, Alejandro Santos-Díaz, and Raquel Valdés-Cristerna

Abstract Heart failure is defined as a syndrome characterized by a remodeling in the cardiac muscle structure that diminishes the strength and synchrony of contractions, leading the subject to a functional capacity deterioration that progressively triggers fatal clinical outcomes. There are specific pharmacological therapies to reduce failure progression. However, once the symptoms have developed, it is necessary to combine different treatments to alleviate them and to improve the quality of life. The most advanced therapies in recent years utilize implantable electronic devices known as cardiac resynchronizers. Cardiac resynchronization therapy has proven to be highly beneficial to patients with severe heart failure, because it reduces the number of hospitalizations, increases exercise resistance, and improves left ventricle systolic function and patient's survival. Nonetheless, 20–30% of patients do not respond to this therapy, thus requiring reliable techniques to predict the probability of a successful outcome in a case-by-case basis. This chapter analyzes several methods that have been proposed to achieve this goal.

The first section reviews the anatomical and physiological basis of cardiac contraction, as well as the main techniques to evaluate it. The second section describes an overview of the medical imaging modalities most frequently used for assessment of the cardiac contraction pattern. Methods of dynamic cardiac image processing such as Fourier analysis and factor analysis of dynamic structures are detailed in the third section, in order to explain statistical models of cardiac

L. Jiménez-Ángeles
Departamento de Ingeniería en Sistemas Biomédicos, Universidad Nacional Autónoma de México, Mexico City, Mexico

V. Medina-Bañuelos · R. Valdés-Cristerna (✉)
Electrical Engineering Department, Universidad Autónoma Metropolitana-Iztapalapa, Mexico City, Mexico
e-mail: ravc@xanum.uam.mx

A. Santos-Díaz
Departamento de Mecatrónica, Escuela de Ingenieria y Ciencias, Tecnológico de Monterrey, Campus Ciudad de México, Mexico City, Mexico

© Springer Nature Switzerland AG 2020
M. R. Ortiz-Posadas (ed.), *Pattern Recognition Techniques Applied to Biomedical Problems*, STEAM-H: Science, Technology, Engineering, Agriculture, Mathematics & Health, https://doi.org/10.1007/978-3-030-38021-2_7

contraction patterns. A strategy to classify the severity of dyssynchrony is also discussed. Finally, a section of perspectives of the analysis of cardiac contraction patterns is presented.

Keywords Cardiac function assessment · Cardiac imaging analysis · Cardiac dyssynchrony · Factor analysis of dynamic structures (FADS)

1 Cardiac Contraction Dynamics

1.1 Anatomical and Physiological Basis of Cardiac Contraction

The heart (Fig. 1) is composed of four chambers, two of them are known as the "right heart" (right atrium and ventricle) and the other two are the "left heart" (left atrium and ventricle). It also contains four main valves (mitral, tricuspid, aortic, and pulmonary). Usually, atria have thin walls and receive blood from the body, the right

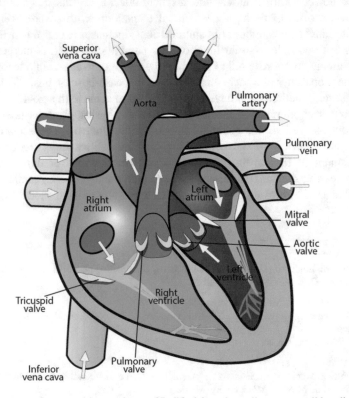

Fig. 1 Anatomy of a normal human heart. (Modified from https://commons.wikimedia.org/wiki/File:Diagram_of_the_human_heart_(cropped).svg)

heart from the bloodstream, and the left part from the lungs. Ventricles are the bigger chambers whose job is to pump blood to the lungs and the circulatory system. The opening of the mitral and tricuspid valves allow filling of the ventricles with blood accumulated in the atria. This phase is known as onset of the ventricular diastole. Once full, they contract to pump the blood to the body in the onset of the ventricular systole.

The series of events happening from the beginning of a contraction to the beginning of the next is known as cardiac cycle. Each cycle is triggered by an action potential that happens spontaneously in the sinoatrial (SA) node, rapidly travels through the atria, and reaches the ventricles via the atrioventricular (AV) node. Due to the composition of the electric conduction system going from the atria to the ventricles, there is a delay in the cardiac stimulus that allows the atria to contract before the ventricles, filling them to their maximum capacity just before the beginning of the ventricular contraction [40].

1.1.1 Electric Conduction System

The heart is equipped with a specialized electrogenic system to create electric pulses that produce the cardiac muscle contraction (Fig. 2). The cells in the SA node have the ability to spontaneously generate action potentials derived from the exchange in sodium, calcium, and potassium ions at the cell membrane level, causing thus an electric discharge followed by an automatic contraction [61, 93]. Hence, the SA node usually controls the beating of the heart.

The SA node is directly connected to the atria muscle fibers in such a way that the action potentials travel outward through the atria muscle mass and reach the AV node.

The conduction system is organized in such way that the cardiac stimulus passes from the atria to the ventricles with approximately one tenth of a second delay that allows the atria to empty their content into the ventricles, before they begin to contract. The AV node and adjacent fibers are mainly responsible for such delay.

Purkinje fibers conduct the electric impulse from the AV node to the ventricles through the bundle of His. Such fibers are considered big and transmit the action potentials at speeds of 1.4–4 m/s, which is six times faster than the ventricular muscle. Such characteristic allows the almost immediate transmission of the cardiac pulse to the rest of the ventricles. A special feature of the bundle of His is its inability to conduct action potentials backward, meaning from the ventricles to the atria. After passing through the fibrillar tissue between atrial and ventricular muscles, the distal portion of the bundle of His goes downward along the interventricular septum for 5–15 mm toward the heart apex. Afterward, it divides into the left and right branches located under the endocardium, in the ventricular septum walls. Each branch extends through its ventricle and divides progressively into smaller branches that round the ventricular cavity and turn toward the heart base. The end of the

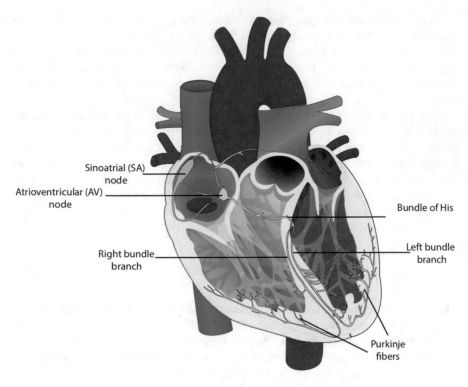

Fig. 2 Electric conduction system of the heart. (Modified from https://en.wikipedia.org/wiki/File: Heart_diagram-en.svg)

Purkinje fibers permeates approximately one-third of the muscle mass thickness where they connect with the muscle fibers. Once the electric impulse reaches the end of the Purkinje fibers, it disseminates to the ventricles through its muscular threads.

1.1.2 Mechanical Contraction

The cardiac muscle constitutes a double-spiral shape with fiber blocks between the spiral layers; thus, the electric impulse does not necessarily travels directly to the heart surface but rather travels with an angle following the spirals [11, 93].

Figure 3 depicts the hemodynamic changes in a single cardiac cycle. During the mechanical contraction of the ventricles, considerable amounts of blood cumulate in the atria due to the fact that the AV valves remain closed. Therefore, when the systole is finished and the pressure in the ventricles drops, the slightly higher pressure in the atria causes the AV valves to open, allowing the blood to rapidly flow into the ventricles; this process is known as the *period of rapid ventricular filling*. Such period lasts approximately one third of the diastole. Immediately after the onset of

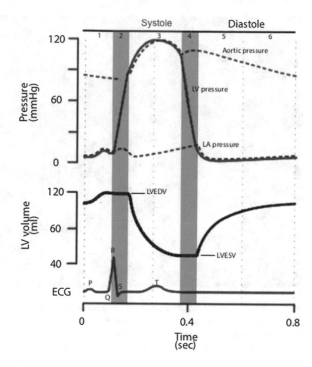

Fig. 3 Cardiac contraction dynamics. Pressure and volume changes during a single cardiac cycle are shown. Pressures are measured in the aorta and left ventricle. Volume curve represents the left ventricle. LVEDV left ventricle end diastolic volume, LVESV left ventricle end systolic volume. (Modified from https://commons.wikimedia.org/wiki/File:Wiggers_Diagram.png)

the ventricular contraction, the pressure inside the ventricles rises abruptly causing the AV valves to close. Afterward, there is a period of 20–30 milliseconds known as *isovolumetric contraction*, where the ventricles contract without emptying, thereby causing the pressure to raise to the point where the sigmoid valves (aortic and pulmonary) open.

When the pressures in the left and right ventricles (RV) rise above 80 and 8 mmHg, respectively, the sigmoid valves open and the blood is ejected. The 70% of the emptying happens during the first third of the systole, whereas the remaining 30% during the last two thirds. These are known as rapid and slow ventricular ejection periods, respectively.

Ventricular relaxation happens rapidly after the end of the systole, decreasing the intraventricular pressures. The elevated pressures in distended arteries push the remaining blood back to the ventricles causing the pulmonary and aortic valves to close. Ventricular muscle continues to relax for another 30–60 milliseconds in a period known as *isovolumetric relaxation* of the ventricles, in which the pressure drops to low diastolic values. Right after, the AV valves open to begin the pumping cycle again.

Chronology of electric events through the conduction system has been described having as a starting point the interventricular septum, extending first to the apex, then spreading to the ventricles and finally to the base of the heart. This has been denominated apex-to-base activation.

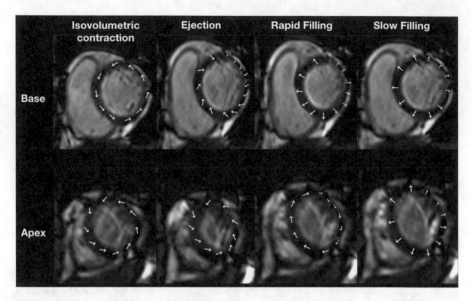

Fig. 4 Cardiac magnetic resonance images showing the phases of the cardiac cycle. Arrows depict the clockwise and counterclockwise directions of transmural twisting motion in a short-axis view, during isovolumetric contraction, mid-systole, isovolumetric relaxation, slower filling, and mid-diastole. (Image modified from Buckberg et al. [18])

In a deeper description of the ventricular contraction mechanics [11, 18], it has been proposed that cardiac muscle can be considered a single muscular band that conforms to a double-loop helicoid, whose fibers extend from the pulmonary artery to the aorta. Such anatomy and a sequential contraction of the band following the muscle fibers trajectory have been suggested as an explanation of the complex motion of ventricles, seen in magnetic resonance imaging studies (Fig. 4) [18]:

(a) A systolic downward displacement of the base of the ventricles toward a motionless apex
(b) An initial systolic clockwise motion of the base and apex, and a change to an anticlockwise motion of the base while the apex continues to move clockwise
(c) Undoing of the previous movements in diastole

1.2 Methods for Evaluation of Ventricular Dynamics

Clinical practice uses multiple tools for cardioventricular dynamic assessment to confirm or reject an illness diagnosis. The evaluation methods described below are the most often considered for diagnosis. However, it is relevant to acknowledge that the selection of a specific technique is a personal decision of the physician.

1.2.1 Electrocardiogram

The electrocardiogram (ECG) is usually the first clinical assessment tool, as it often presents changes in the recording when there are abnormalities in the cardiac conduction system. In particular, approximately 30% of the patients suffering from heart failure (HF) have, in addition to reduced contractility, some defect in the conduction pathway that results in a delay of the systole's onset [36]. Such mismatch may be diagnosed with a duration >120 ms of the QRS complex in the ECG. A delay of the interventricular contraction may complicate further the heart's ability to eject blood. Such delay is associated with clinical instability and high risk of death in patients with HF diagnosis [8, 10, 82, 96].

One of ECG's main features includes its predictive value when abnormal; however, when it shows a normal trace, especially in the parameters related to systolic function, the presence of HF has a probability <10% [29].

Guidelines from the American College of Cardiology, American Heart Association and the North American Society of Pacing and Electrophysiology have approved using implantable devices in order to restore ventricular contraction dynamics and consider duration of the QRS >130 ms as an inclusion criterion, among others, for the prescription of these types of devices [83, 84].

1.2.2 Imaging Modalities

Multiple medical imaging modalities are currently used in the assessment of cardiac function, meaning quantification of anatomical properties (geometry, mass, and volume), mobility of the walls, and assessment of the valves. All of them contribute with valuable information about the nature of a cardiomyopathy and the course of treatment [26]. Figure 5 shows a relative comparison of characteristics for the evaluation of left ventricle (LV) function, using different medical imaging modalities [55].

Measurement of the left ventricle ejection fraction (LVEF), defined as the ratio between the diastolic and systolic volumes [88], is probably the most practical test due to its capability to differentiate patients suffering from systolic malfunction. However, its diagnostic value is somewhat arbitrary. While its normality threshold has been established as >45%, it does not accurately reflect the indices of contractility because its magnitude strongly depends on preload/afterload volumes, heart rate, and cardiac valve function [2, 56, 70]. Thus, cardiac dilation and increased ventricular volumes may keep the LVEF within a range of normal values.

Assessment of the left ventricle's function is fundamental for clinical diagnosis, risk stratification, therapeutic management, and prognosis of patients suffering from cardiac illness. An incorrect evaluation of the left ventricular function may reduce the benefits of a selected treatment or convert it into a high-cost procedure.

Multiple medical imaging modalities such as echocardiography, nuclear cardiology (such as equilibrium radionuclide angiocardiography or ERNA and gated single-photon emission computed tomography or gated-SPECT), cardiac magnetic

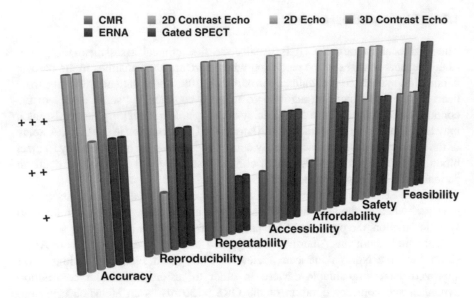

Fig. 5 Relative comparison of characteristics for the assessment of LV function using different noninvasive imaging modalities [55]. CMR cardiac magnetic resonance, ERNA equilibrium radionuclide angiocardiography

resonance (CMR), and multi-slice computed tomography (MSCT) have been widely used and validated in the assessment of ventricular function. Nonetheless, the ideal technique should be accessible, precise, reproducible, safe, and less expensive [55].

The most common modality to evaluate the left ventricular function is two-dimensional (2D) echocardiography [47, 57]. Its widespread use may be attributed to features such as its portability, safety (no ionizing radiation needed), easy procedure, and low cost. However, it is strongly limited by its low reproducibility to determine the LVEF and myocardial wall thickness [45, 46]. Recently proposed, the use of contrast agents may improve the reproducibility of LVEF's evaluation [47].

Development of three-dimensional (3D) echocardiography systems has reduced the limitations of its 2D predecessor, as the LVEF is no longer based on assumptions of the LV morphology. Also, its acquisition time is roughly 3 minutes, considerably faster than other imaging modalities such as CMR [59, 68]. Furthermore, assessment of the ventricular function using 3D echocardiography has been enhanced by the use of contrast agents [63].

ERNA is a well-established modality for assessing LVEF, also widely used in oncologic patients to determine cardiac toxicity due to chemotherapeutic agents [80, 94]. However, a multicenter study using echocardiography, ERNA, and CMR reported that LVEF values measured by these techniques are not interchangeable and that ERNA sub-estimates its magnitude, particularly in patients with a LVEF <35% [15]. Also, it is well known that its main limitations are the overlapping of cardiac structures and the necessity for a background activity correction [94].

The evolution of SPECT systems has allowed them not only to assess myocardial perfusion but also to become a precise and reproducible technique to measure ventricular volumes and the LVEF, even in the presence of perfusion abnormalities [25, 64]. Nonetheless, its low spatial resolution leads to an overestimation of such parameters in small or hypertrophic hearts [91]. Some variables involved in such limitation may be significant perfusion defects, count density, strong peripheral activity, and the volume/size of the left ventricle [91].

CMR is the noninvasive medical imaging modality considered gold standard for the evaluation of left ventricular function [35]. It has several advantages such as reproducibility, high spatial and temporal resolutions, low intra- and interobserver variabilities, and operator independency. Furthermore, ventricular volumes measured using CMR do not include any geometric considerations. Nevertheless, it also carries strong limitations such as low availability, high cost, and conditioning of use in patients with an implanted device (e.g., pacemaker, resynchronizer).

Initially, cardiac MSCT was used in the assessment of coronary arteries' anatomy. Later, the technological advancement made possible to increase spatial resolution and the number of slices up to 64, thus speeding up the acquisition. LVEF and ventricular volumes measured using MSCT have shown good correlation with other modalities such as echocardiography and CMR [43]. Regrettably, it has also shown LVEF sub-estimation when using the 16 slices mode, mainly attributed to the low spatial resolution [14].

Ionizing radiation exposure and the use of contrast agents that may be nephrotoxic for some patients are the main limitations of MSCT. Also, drugs that may not be tolerated by all patients are often used to induce low heart rate values, required to avoid overestimation of the LV peak systolic volume. Furthermore, in addition to a long image post-processing time, a full MSCT study requires a radiation dose of approximately 21.4 mSv [87].

In conclusion, there are many imaging modalities available for the assessment of ventricular function and integrity that will be described in detail in Sect. 2. The ideal method will have to be accessible, reproducible, precise, and safe. Nevertheless, it is worth mentioning that imaging methods and their analysis procedures are not interchangeable; hence, the modality selected to evaluate a particular patient should be used consistently.

1.3   Heart Failure and Cardiac Resynchronization Therapy

Heart failure has a broad definition that accounts for multiple aspects of a complex pathophysiological mechanism directly related with a stimulus, precursor of cardiac muscle damage. Such stimulus unleashes a series of events at a molecular level, which translate into cellular responses that lead to anatomical reshaping of the heart [79]. Thus, HF is defined as a syndrome characterized by a remodeling in the cardiac muscle structure that diminishes the strength and synchrony of the contraction, taking the subject to a functional capacity deterioration that progressively triggers fatal clinical outcomes [31, 61].

During the remodeling of cardiac muscle tissue, there is a progressive deterioration of the heart in its structure and function, as well as the emergence of arrhythmias, which are the main cause of morbidity and mortality among this population [61]. It has been reported that 30% of patients suffering from severe HF show electrical conduction disorders that cause ventricular contraction dyssynchrony, thus increasing the malfunction of the left ventricle [36].

Currently, there is not a unique criterion for the diagnosis of heart failure. Among the more often used, the Framingham criteria consider that HF diagnosis requires the patient to fulfill two major or one major and two minor criteria [58, 81], as shown in Table 1.

The most commonly used classification to quantify the degree of functional limitation caused by the HF was initially developed by the New York Heart Association (NYHA) [54]. Such classification system rated from I to IV the severity of the symptoms when performing a task. Table 2 summarizes the main characteristics of the functional classes.

Functional classification holds an important prognostic value, and it is used as decision criteria in the selection of determined therapeutic interventions, both in medical and surgical cases. Periodic evaluation of functional class facilitates the assessment of the evolution and response to a specific therapy or treatment [48].

The most traditional model of HF assumes a succession of events where abnormalities in the cardiac conductive tissue derive in a delay in the electric pulse conduction through the heart. This generates a delay in the ventricular contraction, a reduction in the cardiac output, and a dilation of the ventricular cavities to

Table 1 Framingham criteria for congestive heart failure

Major	Paroxysmal nocturnal dyspnea
	Orthopnea
	Elevated pressure in jugular artery
	Radiographic cardiomegaly
	Acute pulmonary edema
	Weight loss >4.5 Kg in 5 days with HF treatment
Minor	Bilateral ankle edema
	Nocturnal cough
	Dyspnea on ordinary exertion
	Pleural effusion
	Tachycardia (rate >120 bpm)

Table 2 Heart failure functional classification of the New York Heart Association

Class I	Ordinary physical activity does not cause symptoms
Class II	Ordinary physical activity may be slightly limited by symptoms, but no symptoms at rest
Class III	Physical activity is markedly limited because of symptoms
Class IV	Physical activity cannot be carried out without symptoms; symptoms occur at rest

compensate the loss in ejected blood volume. All these features contribute to HF progression, which means that their correction may benefit the clinical outcome [20, 67, 71].

There are specific pharmacological therapies to diminish the progression of HF; however, once the symptoms have developed, it is necessary to combine different treatments to alleviate them and improve the quality of life. The most developed therapies in recent years utilize implantable electric devices known as cardiac resynchronizers [84].

Cardiac resynchronization therapy (CRT) has shown to be beneficial and improve the functional class of patients with severe HF. It also reduces the number of hospitalizations, increases exercise resistance, and improves left ventricle systolic function and patient survival [3, 4, 28]. Nonetheless, 20–30% of patients do not respond to CRT [27]. Also, it has been suggested that in order to have a positive outcome from CRT, it is necessary to consider the mechanical contraction dyssynchrony, which can be present as atrioventricular, interventricular, or intraventricular dyssynchrony. Regrettably, it is currently not known which of these plays the main role for a positive CRT response [27], but the loss in mechanical contraction synchrony (dyssynchrony) has been proposed as a reliable response indicator [97]. This has led to a continuous search for a method capable of accurately measuring the mechanical dyssynchrony of the ventricles and that can be considered a predictor for CRT outcome.

Parallel to the development of treatments for HF, different medical imaging modalities have been used to assess the impact of CRT in the ventricular contraction mechanics restoration as well as to improve patient selection of those whom will benefit from it. However, it has been suggested that imaging modalities such as the echocardiogram have a low reproducibility, sensitivity, and specificity to detect the patients that would benefit from a CRT [98].

2 Cardiac Imaging for the Assessment of Contraction Patterns

2.1 Echocardiography

2.1.1 Conventional M-Mode

Cardiac contraction mechanics can be determined using M-mode ultrasound images. A first approach consists on measuring the delay between the motion of the septum and the posterior wall of the heart; a time period longer than 130 ms for this parameter reflects severe intraventricular dyssynchrony (Fig. 6). This method makes also feasible to determine the contraction of the lateral wall and measure its delay with regard to the onset of the diastolic filling: simultaneous appearance of systolic contraction and diastolic relaxation is also indicative of severe intraventricular dyssynchrony [39, 42].

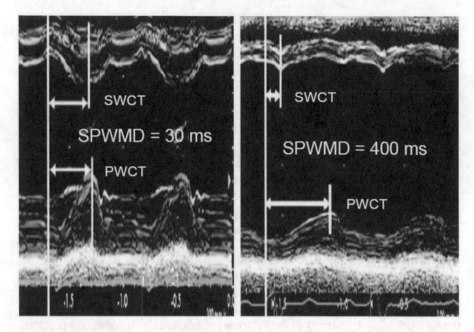

Fig. 6 Determination of septal to posterior wall motion delay (SPWMD) from M-mode imaging in a normal subject (left) and in a patient with congestive heart failure (right). SPWMD corresponds to time difference between septal (SWCT) and posterior (PWCT) wall contraction times. (Adapted from Galderisi et al. [39] with permission)

These measurements show high specificity but low sensitivity. Additionally, there are some disadvantages when determining them: they are very sensitive to the incidence angle of the acoustic waves on the myocardium, as ideally the transducer should be placed perpendicular to the wall of the left ventricle, and it is often hard to find the access window. Additionally, motion of the septum and the posterior wall are frequently diminished or absent in patients suffering from ischemia. Hence, the measurement reflects not only intraventricular but also interventricular dyssynchrony, making its validation and interpretation more difficult [12, 42]. These problems are common among echocardiographic techniques.

2.1.2 Tissue Doppler Imaging

Doppler ultrasound has been widely used in medical applications for the measurement of blood flow. A particular technique of the method, known as tissue Doppler imaging (TDI), consists on highlighting the signal produced by myocardial motion, which is usually slower but of higher amplitude compared to that of blood flow (myocardium velocities are roughly 10% compared to blood flow). Hence, it is possible to measure the velocity of a selected point on the myocardium as it gets

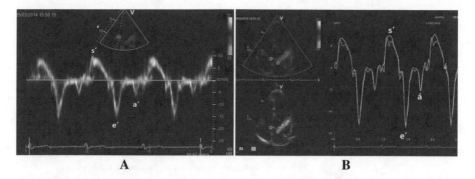

Fig. 7 Pulsed-wave and color-coded Doppler tissue imaging. For illustration purposes, systolic (s') and diastolic (e', a') velocity of the aortic annulus (AA) are presented. (**a**) Pulsed-wave Doppler tissue imaging at the septal site of AA. (**b**) Velocities measured at two sites (septal and lateral) of AA by using quantitative two-dimensional color Doppler tissue imaging. (Adapted from Nygren et al. [65] with permission)

closer or further from the transducer and also to determine the displacement or absolute distance through the integration of the velocity curve over a time period.

TDI can be performed using either pulsed Doppler (Fig. 7a) or colored Doppler (Fig. 7b). The former is widely available, is acquired at high temporal resolution, and allows real-time analysis. However, due to the fact that only one point can be selected during a single acquisition, it has a low spatial resolution and it is impossible to reposition the sample or to compare different segments. Furthermore, it has been reported that velocities determined using this method are 20% to 30% higher when compared with color-coded TDI [5]. On the other hand, the later allows simultaneous analysis of multiple points and comparison of their contraction patterns, resulting in a faster acquisition with higher spatial resolution. Nonetheless, its availability is limited and the analysis must be performed in off-line post-processing. A better analysis of ventricular mechanics may be achieved using a single acquisition of three orthogonal planes that allow comparing the 12 segments of the heart simultaneously, during the same cardiac cycle [42] and the repositioning of the sample in other orientations.

Velocities of interest on TDI correspond to the longitudinal motion seen on the apical projections of two, three, or four chambers. In such direction, the apex remains motionless, whereas the base approaches it during systole and moves away during diastole. The motion difference between these anatomical regions generates a velocity gradient as the base shows the highest value, whereas the velocity in the apex remains almost unchanged, as observed in Fig. 8 [5].

In order to establish a dyssynchrony index, multiple segments of the myocardium must be analyzed according to standards from the American Heart Association and the American Society of Echocardiography [22]. Most reported studies include 12 positions: 6 basal and 6 central segments of the left ventricle. The periods of interest are the peak systolic velocity (T_s) or the onset of systolic velocity (T_0), measured from the beginning of the QRS wave in the ECG. In normal conditions during the

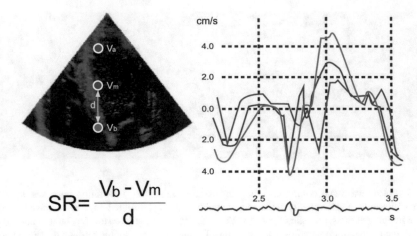

$$SR= \frac{V_b - V_m}{d}$$

Fig. 8 Measurement of strain rate. Normal heart motion is such that the base moves toward the apex, which remains almost motionless. Thus, tissue velocity reaches a maximum at the base (V_b), becomes lower in the mid-heart (V_m), and becomes slowest at the apex (V_a). The strain rate is calculated as the difference between velocities measured at two points divided by the distance between them (d)

systole, all segments move toward the transducer and show their peak values at the same time. On the other hand, during diastole, the segments move away from the transducer and produce myocardial relaxation velocity curves.

The measures that best reflect the ventricular dyssynchrony are the standard deviation of the 12 segments or the maximum time difference between some of them. Ventricular dyssynchrony can be established when there is a time difference higher than 65 ms in the peak systolic velocity of the 12 segments and a standard deviation cutoff of 34 ms in the peak value of their curves [12, 85].

Nevertheless, this method holds some disadvantages: TDI explores the motion of specific segments of the heart and is not capable to completely assess the myocardial mechanics; it is difficult to determine the peak systolic velocities and the standard deviation for all segments; it also requires an appropriate transducer orientation and a good location of the exploration window [5, 12].

2.1.3 Tissue Synchronization Imaging

The tissue synchronization imaging (TSI) method automatically detects the peak systolic velocity, as it uses information from TDI to assign color to the myocardium as a function of each segment's peak velocity. These images are superimposed on real-time images (Fig. 9) that allow detecting regional delays, from the early peaks to those showing elongated times. The mathematical median of the T_s values from a volume sample manually selected allows comparing different segments in a quantitative way and shows a "snapshot" of the presence or absence of

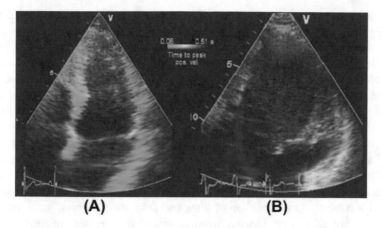

Fig. 9 Tissue synchronization images where time-to-peak velocities are color-coded as follows: green (20–150 ms, normal), yellow (150–300 ms), and red (300–500 ms, dyssynchrony). (**a**) Healthy subject and (**b**) patient with left bundle branch block, where an abnormal contraction in basal and mid-septal segments is shown. (Reproduced from Knebel et al. [53] with permission)

dyssynchrony. However, it is necessary to manually select the onset and end of systole by selecting the opening and closing times of the aortic valve, making the method susceptible to timing errors. Additionally, it has a lower sensitivity compared with other methods [12, 42].

TSI also allows measurement of tissue deformation or tissue strain, defined as the normalized change in length with reference to the original length, as well as the velocity of such change, known as strain rate. In clinical cardiology, the peak systolic strain rate is the parameter that best describes the local contractile function. Contrarily, tissue strain is dependent on the volume and thus shows a poor description of contractility. Taken from TDI, the strain rate allows calculating the myocardial wall velocity gradient by measuring the difference between the velocities in two points. A velocity gradient can also be obtained between the endocardium and epicardium, as well as the base and apex torsion motions across the cardiac cycle. Tissue strain and strain rate are less susceptible to translation and tethering motion, thus better describe the myocardial function when compared with tissue velocity measurements [5].

2.1.4 Speckle Tracking

In order to suppress the angular dependency of TDI, another echocardiographic method to determine tissue deformation has been proposed, using conventional B-mode images. It consists of tracking the acoustic reflections and interference patterns within an acoustic window. These phenomena create speckle markers distinctive of myocardial tissue. Such markers associate to each other in a frame-by-frame basis using a simple addition of absolute differences, to generate deformation and tissue motion sequences in both 2D and 3D [41]. This method is capable of creating strain maps in the radial, circumferential, and longitudinal directions.

Speckle tracking analysis is performed using standard bidimensional images acquired in apical views of two and four chambers, as well as a short-axis parasternal view taken at the papillary muscles level, looking for the most circular projection of the heart. A region of interest (ROI) containing stable speckle patterns is determined manually and subdivided into smaller windows of 20–40 pixels, in order to perform the analysis across the temporal sequence. A minimum frequency of 30 frames per second is recommended, where the selection of the frame corresponding to the end of systole will serve as reference to tag the ventricular cavity, making sure of including all the segments to be explored. Then, speckle patterns are followed on the subsequent frames across the entire cardiac cycle (Fig. 10). Distance between these patterns is measured as a function of time, and from those curves, the deformation parameters are calculated. Finally, the segments of interest are assigned a color in accordance with the measured strain, and the curves for radial (RS), circular (CS), and longitudinal (LS) strain are displayed (Fig. 11) [30].

From the strain maps mentioned above, ventricular dyssynchrony can be determined using two parameters:

- Using CS and RS, the difference between time-to-peak systolic strain of the septal and posterior segments is measured, as well as the standard deviation of time-to-peak systolic strain for the six segments in the short-axis projection.

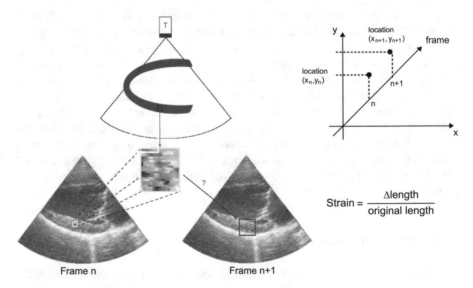

Fig. 10 Speckle tracking applied on a region of interest. The relative position of the speckle pattern on a frame-by-frame basis is used to measure tissue motion. Strain is determined as the percentage of longitude change. (Adapted from Jasaityte and D'Hooge [49] with permission, Imaging in Medicine ©)

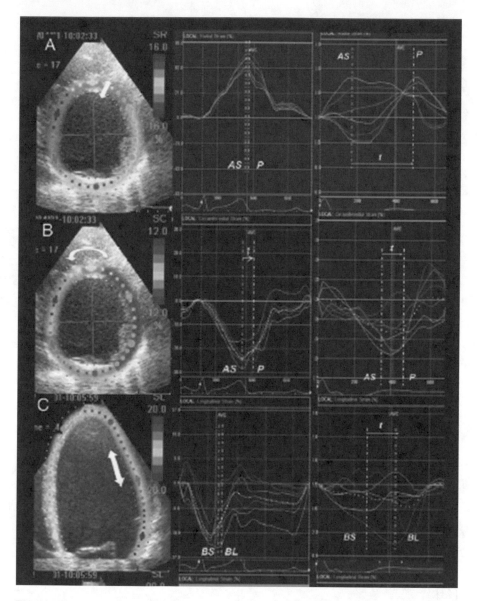

Fig. 11 2D strain images. (**a**) Radial thickening; (**b**) circumferential shortening; (**c**) longitudinal shortening. Middle and right columns show segmental time-strain curves for synchronous and dyssynchronous contractions, respectively. (Reproduced from Delgado et al. [30] with permission)

- For LS, the two- and four-chamber images acquired are used to calculate the difference between time-to-peak systolic strain of the basal-septal and basal-lateral left ventricle segments and the standard deviation of time-to-peak systolic strain for the 12 segments in the left ventricle [30, 85].

Radial strain maps have shown superiority over longitudinal measurements when determining dyssynchrony and correlate well with indices obtained using TDI. Another advantage of this method is its non-dependency of the incidence angle, which makes it ideal in complicated access sites for TDI. However, it requires high-quality images acquired at a high frame rate, and its temporal resolution is lower in comparison with tissue Doppler. It also requires an accurate follow-up of stable speckle patterns, which makes manual adjustment of the ROIs often necessary [42].

2.2 Cardiac Radionuclide Imaging

2.2.1 Equilibrium Radionuclide Angiocardiography

Nuclear cardiology methods hold some advantages over echocardiography as they are less operator dependent and highly reproducible and can be easily automated. They are based on the fact that normal myocardial tissue contracts in a coordinated fashion, such that most of ventricular and auricular segments are almost in phase [60].

Since the early days of nuclear cardiology, evaluation of ventricular dynamics has always been the focus of several methods. In 1971, the pioneers Zaret and Strauss proposed a method to synchronize a tagged blood bolus with the ECG signal. This was the precursor of ERNA for the determination of left ventricle ejection fraction (LVEF) and detection of abnormal myocardial motion. Soon after, Borer et al. combined ERNA with an exercise challenge, allowing global and regional assessment of left ventricle function under stress. The tomographic methods that appeared during the 1980s and 1990s pushed forward the myocardial perfusion analysis; techniques such as SPECT and positron emission tomography have proven their clinical utility for the assessment of ventricular function [92].

The ERNA study is a set of images that represent the spatial distribution of a radiotracer (usually technetium-99 metastable (Tc-99 m) bonded to erythrocytes) at a specific moment of the cardiac cycle and relate pixel's intensity to ventricular volume. Figure 12a shows the 16 frames of an ERNA planar study, acquired using a left anterior oblique (LAO) view. Additionally, it shows the ROIs corresponding to ventricles at the end of diastole (frame 1) and systole (frame 8), respectively. Figure 12b shows the nomenclature for the left ventricle walls according to Hesse et al. [44].

The LAO view is frequently used in the evaluation of ventricular function due to the fact that it shows a clear definition of both ventricular cavities with a noticeable separation in the interventricular septum. However, a limitation of this view is that the atrial cavities are partially hidden behind the ventricles. In practice, atria can be better evaluated when sequentially showing the series of frames, looking for changes in signal intensity, with respect to ventricles. Also, abnormalities happening in the output tract of the right ventricle and the aorta can be identified. Nevertheless, a

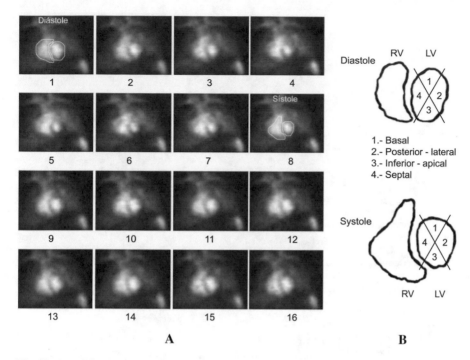

Fig. 12 (**a**) 16-frame planar ERNA study in LAO view. (**b**) Nomenclature of the left ventricle walls

good definition of cardiac cavities and structures will depend upon the erythrocyte labeling method and image acquisition parameters.

Blood labeling can be performed in two ways: for in vitro labeling (frequently used), 5 ml of blood is extracted from the patient, mixed with pyrophosphates, labeled using 30–35 mCi of Tc-99 m, and finally reinjected intravenously. Nowadays, a commercial erythrocyte-labeling package UltraTag™ RBC Kit (Mallinckrodt Nuclear Medicine, 1995, St. Louis, MO, USA) can achieve an efficiency of 95%. On the other hand, for in vivo labeling, a bolus of a stannous pyrophosphate is injected directly to the patient which, after circulating for approximately 30 minutes, binds to erythrocytes that are then labeled through an intravenous injection of 30–35 mCi of Tc-99 m and distributed uniformly in the blood stream, achieving an efficiency of 90% [74].

Acquisition of the study requires the use of a gamma camera capable of detecting the γ-rays coming from the radiotracer previously injected to the patient. Figure 13 shows the detection process and conversion of radiation.

Image acquisition is synchronized with the R wave of the ECG, positioning the detector of the gamma camera on the patient in a supine position, using the LAO view, between 30° and 45° to maximize visualization of the septum and separation of the ventricles. A 64 × 64 pixel acquisition matrix is commonly used with a zoom factor of roughly 1.3 to easily visualize the heart at the center of the field of view and isolate other structures such as the liver and the spleen.

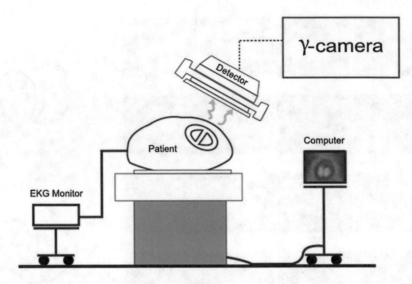

Fig. 13 Schematics of the ERNA study acquisition

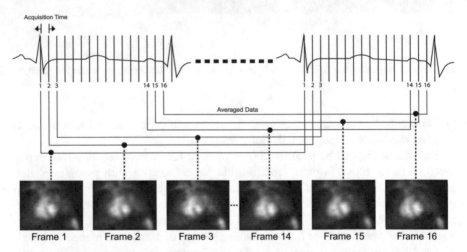

Fig. 14 Acquisition principle of ERNA studies. R-R interval in the ECG trace is divided evenly in 16 segments. Data are acquired for each interval along several cardiac cycles, and resulting images are a weighted average of each time interval

The temporal resolution of the study is a function of the number of frames used to describe the entire cardiac cycle (usually 16 or 32). Gated acquisition with the R wave allows building up series of images across several heart beatings, each frame representing a specific moment of the cardiac cycle. Due to heart rate variability, an acquisition window (usually 10%) is established to rule out heart beatings with a different duration. Figure 14 depicts the acquisition principle of ERNA studies,

Table 3 Acquisition parameters for ERNA studies according to the American Society for Nuclear Cardiology [52]

Parameters	Recommendation
Radiotracer	Erythrocytes labeled using Tc-99 m
Dose	20–25 mCi/70 kg
Labeling method	In vivo, in vitro, in vivo/in vitro
Pixel size	2–3 mm/pixel
Frames	16–32 frames/cycle
Count density	20,000/cm^2

whereas Table 3 summarizes its parameters according to the American Society of Nuclear Cardiology [52].

ECG gating can be performed in three different ways [44]:

(a) *Forward gating*: The R wave triggers acquisition distributed in time intervals defined by the average duration of the cardiac cycle and the number of images in the study. The first frames (systole) are usually more accurate than the last (diastole) due to heart rate variability.

(b) *Backward gating*: The R wave of the following cardiac cycle is used as reference; meanwhile, the counts are distributed in time intervals retrospectively. In this method, intervals corresponding to diastole are more accurate than those of systole.

(c) *Variable Timing*: The time interval is calculated individually for every cardiac cycle. This method allows better timing, but it is not widely available.

The end of acquisition can be programmed based on the number of cardiac cycles acquired, the total number of counts, or a time period. The criteria that assures best image quality is the total number of counts, which indicates that an image of optimum quality is built with 250,000 counts, adding up to eight million counts per study of 32 frames [44]. The duration of a 32 frames study is usually between 10 and 15 minutes.

In order to analyze ventricular function, left and right ventricles are defined (manually or automatically) in each frame, and time-activity curves (TACs) are built and analyzed. These represent the temporal evolution of signal intensity across the entire cardiac cycle. Due to the fact that the radiotracer is uniformly distributed in the bloodstream, variations in signal intensity are proportional to blood volume. Thus, the plot of activity within a ventricle during an entire cycle is known as *ventricular volume curve*.

Equilibrium radionuclide angiocardiography can be obtained in a planar way as described above or in combination with a tomographic method such as SPECT. ERNA-SPECT studies use the tomographic reconstruction principle in order to produce series of bidimensional images of adjacent slices of an organ or tissue of interest. ERNA-SPECT studies are acquired over a 180° arc from the left posterior oblique (LPO) to the right anterior oblique (RAO) views generating a 4D dataset: 3D images corresponding to the short, long vertical, and long horizontal axes

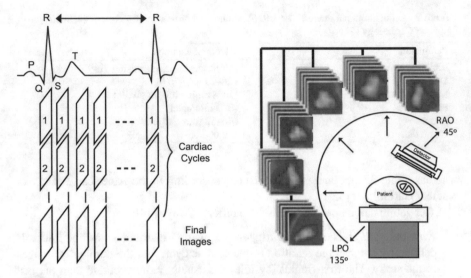

Fig. 15 Schematics of the ECG-gated image acquisition: cardiac cycle is divided evenly in a number of frames; each resulting image is a weighted average from several cardiac cycles. An arc of 180° is used to acquire an ERNA-SPECT study. RAO right anterior oblique view, LPO left posterior oblique view

projections, each with several slices using ECG gating [72]. Figure 15 depicts the gated ERNA-SPECT image acquisition principle [74].

Both methods (planar ERNA and ERNA-SPECT) allowed for development of parametric representations, based on digital processing of the TAC curves, such as amplitude and phase images. The former represents the magnitude of the contraction, whereas the phase image depicts a map of the ventricular contraction sequence; both are used in clinical practice for the evaluation of cardiac dyssynchrony. These methods are described in detail in Sect. 3.

2.2.2 Gated Myocardial Perfusion Single-Photon Emission Computed Tomography

ERNA studies explore multiple frames of the ventricular function per cardiac cycle, thus achieving a high temporal resolution. However, it only shows 2D planar views of oblique projections. This limitation is overcome using synchronized tomography in a method called gated myocardial perfusion SPECT (GMPS), which has been widely used in recent years. Contrary to ERNA, GMPS is based on determining the myocardium wall thickening time during a cardiac cycle, depicted by brighter areas in the systole. Hence, time-activity curves in GMPS correspond to a tissue thickness curve, with a reduced temporal resolution of only 8 or 16 sample points per heart cycle. A typical GMPS study gathers information from many segments (over 600), determines the onset or contraction peak, and compares it between segments. There are many commercial systems offering automated software to perform this analysis.

GMPS is used to determine parameters indicative of LV dyssynchrony such as phase histogram bandwidth, phase range including 95% of the histogram, and standard deviation of the phase distribution. Other useful parameters include entropy, skewness, and kurtosis [33].

GMPS has some advantages compared to ERNA, as the same study is useful to determine myocardial perfusion, wall motion, dyssynchrony, and ventricular function. Nonetheless, this includes the left ventricle only, whereas ERNA allows assessing interventricular dyssynchrony as well. Thus, the latter is still considered the gold standard in clinical practice.

2.3   Cardiac Magnetic Resonance

Many magnetic resonance imaging techniques have been used for the evaluation of cardiovascular morphology, ventricular function, myocardial perfusion, and tissue characterization. This modality, better known as CMR, requires breath holding, ECG, and respiratory gating in order to avoid motion artifacts. Two methods are commonly used for ECG gating: The first method, *prospective gating*, uses the R wave to trigger the acquisition of a set of k-space lines in a short time period, which defines the temporal resolution. This sequence is repeated in the subsequent R-R intervals until the full k-space is acquired, for image reconstruction [90]. Acquisition time must be inferior to an R-R period, so it is commonly adjusted to 80% or 90% of that value. However, this restricts the acquisition of images at the end of diastole. The second method, *retrospective gating*, acquires data continuously and matches it with the ECG afterward. Due to the fact that data are reassigned to k-space according to their relative position with respect to the ECG, it is possible to get images from systole and diastole [90].

There are multiple CMR techniques for cardiac dyssynchrony assessment. Some examples are 2D and 3D myocardial tagging, velocity-encoding MR, or CMR tissue resynchronization, whose capability to assess successful CRT response has been widely reported. However, CMR's limited availability, high cost, long acquisition times, and restriction of use in patients with implanted devices have restrained its routine use in clinical practice [33].

Not only the appropriate protocol selection for CMR acquisition but also dyssynchrony parameters play a key role in the evaluation of cardiac function. Similarly as in echocardiographic techniques, the most common parameters are based on measuring contraction peaks, time delays between opposite wall motion, or standard deviation of peak strain duration for the 12 ventricular segments [19]. The disadvantages of such parameters (difficulty and subjectivity when identifying velocity peaks, especially in dyskinetic or infracted myocardia) may be overcome by measuring dyssynchrony indices based on longitudinal, radial, and circumferential homogeneity ratios. Such spectral measurements are determined in the Fourier

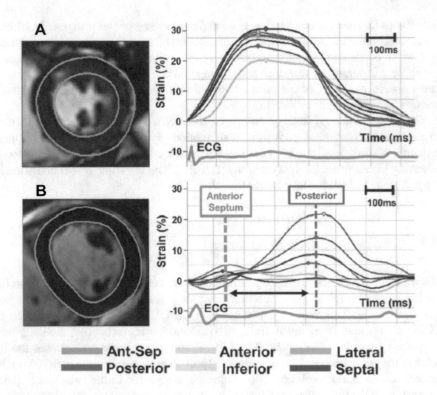

Fig. 16 Radial time-strain curves measured by feature tracking CMR in (**a**) a healthy subject (QRS = 92 ms) and (**b**) a patient with LV dyssynchrony (QRS = 138 ms). (Adapted from Onishi et al. [66] with permission)

domain of the regional strains, thus not requiring validation of velocity peaks or strain curves. Due to the fact that synchronic contraction is represented by zero-order terms of the Fourier transform, whereas dyssynchrony by the first-order terms, the relationship defined by the square root of the quotient between the zero-order term and the addition of the zero- and first-order terms may be used as a reliable dyssynchrony index [19].

As an example of this technique, in Fig. 16a, it can be appreciated a subject with normal left ventricular function shown as synchronous time-to-peak strain curves, measured in six segments. A case presenting ventricular dyssynchrony is shown in Fig. 16b, where delayed time-to-peak-strain curves with early peak strain in anterior septum segment and late peak strain in posterior lateral segments can be observed [66].

CMR offers two main pulse sequence groups for evaluation of cardiac function: First, myocardial tagging creates artificial labels on the image as a mesh of signal voids that are followed during the image acquisition. Such tags appear on the images as black stripes that move with the myocardial segment linked to them. This mesh can be implemented in two ways: selective saturation or modulation of the magnetization vector using magnetic field gradients. Both options are available in commercial systems.

The second approach of cardiac functional imaging is phase-contrast velocity encoding; this method consists of applying bipolar gradients along the desired direction before signal acquisition. Due to the fact that its velocity modulates the phase of the explored object, each pixel contains information of such parameter; hence, high-resolution functional data can be acquired. Nonetheless, only short periods of motion can be registered, and encoding low velocities requires the use of stronger gradients, resulting in long echo times. Additionally, cardiac motion generates phase distortions, especially in the LV posterior wall [6].

Cine MRI is a modality that allows acquiring images of the moving heart across the entire cardiac cycle. As a result of the strong contrast between myocardial tissue and the blood within the cardiac chambers, periodic changes and local motion of ventricular walls can be clearly observed and quantified. Cine MRI studies use a balanced steady-state free precession protocol, consisting of a fast gradient-echo (GRE) pulse sequence where phase shifts induced by the gradients are equal to zero every repetition time. Cine MRI grants acquiring two- and four-chamber and short-axis images of the heart, which is the gold standard for systolic function and ventricular dyssynchrony validation [90].

Phase-contrast cine acquires a set of images through the cardiac cycle in a similar fashion as cine MRI. Phase images are acquired to indicate flow velocity in addition to magnitude images. This is accomplished by applying an additional bipolar velocity-encoding gradient to the GRE sequence in order to measure the phase shift caused by blood flow. This phase shift is proportional to flow velocity in the gradient direction, such that stationary tissue appears in a mid-gray intensity, whereas the blood shows a bright or dark signal depending on the direction of blood flow, referenced to the velocity-encoding gradient [90]. This same principle is also applied in the measurement of myocardial wall motion velocity, considering they are slower than those of blood flow.

In order to alleviate the disadvantages from myocardial tagging and velocity encoding, it is possible to combine both perspectives and acquire high-resolution images where measuring big tissue displacements in long time periods is feasible. Displacement encoding stimulated echo (DENSE) is a method to codify tissue displacement directly from the phase of the MR signal, thus improving contrast with moderate gradient strengths [6].

3   Analysis of Cardiac Contraction Dyssynchrony by ERNA

As described in Sect. 2.2, an ERNA study is a series of images representing the homogeneous spatial distribution of a radiotracer within the ventricular cavities throughout an entire cardiac cycle, and it relates the pixel's intensity with ventricular volume. In this section, the complete procedure of cardiac contraction analysis is described, based on results previously reported by our research group.

3.1 Data Acquisition

The ERNA studies processed in this section were acquired as follows [75, 77]:

Planar ERNA A General Electric Millennium MPR/MPS Gamma Camera was used for all ERNA image acquisition. The camera contains a single head with 64 photomultiplier tubes, and it is equipped with a low-energy high-resolution parallel-hole collimator; the calibration of the energy peak was centered at 140 KeV, and the detector uniformity was guaranteed at less than 5%. Images were digitized at a 64×64 pixel resolution and 1.33 as zoom factor.

Erythrocytes were tagged applying an in vivo/in vitro modified technique with 740–925 MBq of Tc-99 m, using an UltraTag™ RBC Kit (Mallinckrodt Nuclear Medicine, 1995, St. Louis, MO, USA) [13]. Electrocardiogram trace was continuously monitored to synchronize image acquisition with the R wave. To eliminate ventricular extrasystoles during acquisition, a beat acceptance window was defined at +/−20% of the average heart rate. Images were taken in an anterior left oblique projection to attain the best definition of left and right ventricles simultaneously. A total of 16 frames were obtained with a density of 300,000 counts per frame; however, in the analysis, the last frame of each study was eliminated due to the low quality of the image as a result of the R-R interval variability during the acquisition, which leads to a low signal-to-noise ratio.

ERNA-SPECT Red blood cells were tagged as usual for ERNA studies. The data were collected at 32 projections over $180°$ with a frame size of 64×64 pixels, acquired for 50.6 s per projection for 27 minutes. Images with patients at rest were acquired at 16 frames per R–R interval, using an R-wave window of $\pm 70\%$ of mean pre-acquisition heart rate. All images were acquired with the same single-head gamma camera, equipped with a low-energy and high-resolution collimator.

3.2 Fourier Phase Analysis

Let be $X_{TAC}(p, q) = X_{TAC}(p(i, j), q(k))$, a matrix whose rows represent the (i, j)-th pixel value of the k-th frame following the structure $p = (i - 1) \times N + j$, $q = k$, for frames of size $N \times N$ pixels (Fig. 17). From this matrix, $X_{TAC}(p)$ represents the temporal evolution of the pixel p in the ERNA study, known as time-activity curve.

Fourier phase analysis first proposed by Botvinick et al. [16, 17] is the traditional method to analyze cardiac contraction dyssynchrony using ERNA studies. This method assumes that each TAC is periodic; thus, it can be adjusted to the first harmonic of its Fourier transform, resulting in curves characterized by an amplitude and phase. Then, this information is used to build amplitude and phase maps representing all the TACs from the ERNA study.

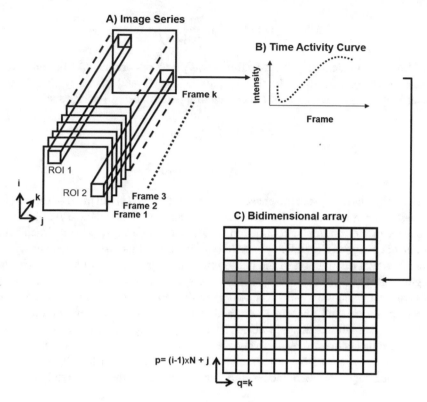

Fig. 17 (**a**) Series of ERNA images with k-frames. (**b**) Time-activity curve for a ROI in the image. (**c**) Bidimensional array X_{TAC}. (Adapted from Santos-Díaz et al. [77] with permission)

Let $x[n]$ and $X[v]$ be two discrete series with the following relationship:

$$x\,[n] \overset{\mathcal{F}}{\leftrightarrow} X\,[v]$$

$x[n]$ is a discrete and periodic series in the time domain, whereas $X[v]$ is discrete Fourier transform, that is, a discrete and periodic series in the frequency domain of the form:

$$X\,[v] = \sum_{n=0}^{K-1} x\,[n]\, e^{\left(\frac{-2\pi j}{K}\right)nv} \quad \text{for } v = 0, 1, \ldots, K-1 \tag{1}$$

From $\mathrm{Re}\{X\}$ and $\mathrm{Im}\{X\}$ of the first harmonic ($v = 1$) in Eq. 1, each TAC has a unique magnitude and phase defined as:

$$A = \sqrt{\mathrm{Re}\,\{X\}^2 + \mathrm{Im}\,\{X\}^2}$$

$$\phi = \tan^{-1}\left(\frac{\text{Im}\{X\}}{\text{Re}\{X\}}\right)$$

Throughout calculating the magnitude and phase of every TAC from an ERNA study, it is possible to build the magnitude (*FoAI*) and phase (*FoPI*) images describing an approximation of the contraction pattern, using Fourier analysis.

$$FoAI\,(i,\,j) = A\,(i,\,j)$$

$$FoPI\,(i,\,j) = \phi\,(i,\,j)$$

where $\phi(i,j)$ and $A(i,j)$ are the phase and amplitude of the (i,j)-th pixel of the ERNA.

FoAI is a reflection of the contraction magnitude, whereas *FoPI* reflects the sequence of the ventricular contraction, coded in a 0–360 degrees cycle. Each value in *FoPI* is assigned a specific color, and both image and its histogram are analyzed. The statistical distribution of the phase image histogram relates its mode with the beginning of the contraction, whereas its dispersion is related to the contraction dyssynchrony. Figure 18 depicts two examples of phase and amplitude images showing normal and abnormal contraction patterns. The left column shows the color-coded amplitude image, whereas the central and right columns show the phase image and its histogram, respectively, the latter using the same color map.

The top row shows the results of a study performed in a healthy subject; its amplitude image shows a homogeneous intensity region in the LV; the right ventricle

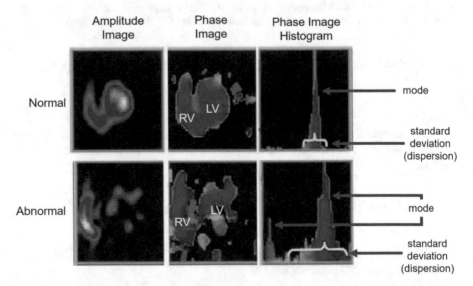

Fig. 18 Amplitude and phase images obtained from the Fourier analysis performed in X_{TAC} arrays of subjects having a normal (top) and abnormal (bottom) cardiac contraction patterns

(RV) shows also an almost homogeneous but lower intensity. Additionally, the phase image histogram shows a low dispersion corresponding to both ventricles, as is expected from a healthy contraction pattern, where both contract simultaneously. On the other hand, the bottom row presents the results for a heart failure patient. The *FoAI* contains high-intensity values in the RV region only, whereas the LV shows just a few low-intensity spots. The *FoPI* histogram has a bimodal distribution with the main mode having a wide dispersion and the second grouping some pixels from the interventricular region. This is a case of inter- and intraventricular dyssynchrony.

Nonetheless, Fourier analysis for modeling of the cardiac contraction pattern holds some limitations:

- The temporal evolution of the pixels' intensity change approximates a periodic function.
- Its morphology can be described by only one harmonic of the Fourier series.

While such considerations approximate the contraction pattern of a healthy subject, this model is insufficient for an abnormal one; however, statistical interpretation of the phase image found clinical value by analyzing its difference with respect to the expected pattern for a normal contraction.

Chen et al. [24] used the Fourier phase analysis to assess its capability for predicting the acute response to CRT in a cohort of heart failure patients, specifically suffering from left bundle branch block (LBBB). The observed parameters were the LV volume and its intraventricular dyssynchrony, characterized by the standard left ventricular phase shift (LVPS) and left ventricular phase standard deviation (LVPSD). Patients were labeled as acute responders or nonresponders based on a 15% reduction of the LV end systolic volume, immediately after the resynchronizer was implanted. ERNA studies were performed pre and post the CRT procedure as well as 48 hours after. The acute responder group showed a significant reduction in the LVPS% and LVPSD% right after CRT, whereas the nonresponders remained with no change. The prediction model established that a 25% threshold in LVPS% reaches a sensitivity and specificity of 80% and 89%, respectively (AUROC $= 0.93$), whereas an 8.5% threshold in LVPSD% showed 87% sensitivity and 89% specificity (AUROC $= 0.95$), when predicting acute response to CRT.

In a similar study, Tournoux et al. [89] assessed the predictive value of Fourier phase image analysis from ERNA studies in CRT response. They included 147 patients with a complete clinical record and ERNA images taken between 2001 and 2011. For the subgroup of patients with implant (57%), images were acquired before the procedure. Long-term follow-up was performed in order to establish the benefits of CRT, registering the cases of death by any cause and heart transplant as non-desirable conditions for prognosis. Parameters included in the analysis, taken from the Fourier phase image histogram were as follows: standard deviation on the left ventricle phases (iLV-SD), the delay between the earliest and most delayed 20% of the LV phases (iLV-20/80), standard deviation of the RV phases (iRV-SD), and the difference between LV and RV (LV-RV) mode phase angles. As expected, CRT patients showed a better prognostic within a 3-year follow-up. A statistically significant difference in the probability of survival was found between the groups of patients with and without implant, only in those cases of ventricular dyssynchrony confirmed by ERNA. For those cases of moderate ventricular dyssynchrony, no difference in the prognostic of survival was found.

3.3 Factor Analysis of Dynamic Structures

The factor analysis of dynamic structures (FADS) takes the group of time series $X_{TAC}(p, q)$ curves and describes the TACs as a superposition of dynamic factors weighted by coefficients that hold the principal time variation modes. The first step of FADS consists of determining a subspace containing the vectors that represent the temporal evolution of the pixels without redundancy. The dimensionality of such subspace is equal to the number of frames in the series (K). These non-correlated vectors are calculated using principal component analysis, and the factors are a result of the Karhunen-Loève transform [86].

The model can be described as follows:

$$X_{TAC}(p, q) = f_{p,1}c_{1,q} + f_{p,2}c_{2,q} + \cdots + f_{p,K}c_{K,q}$$

$$X_{TAC}(p, q) = \sum_{i=1}^{K} f_{p,i}c_{i,q} = \boldsymbol{f}_p * \boldsymbol{c}_q$$

where $\boldsymbol{f}_p$ is the vector of dimension $1 \times K$ that represents the contribution of each dynamic factor to a position p, whereas $\boldsymbol{c}_q$ is the vector of the q-th coefficients from the $K \times K$ temporal variation modes. Its dimensionality is $K \times 1$.

After reconstructing the whole group of time series, it follows that:

$$X_{TAC} = FC \tag{2}$$

where the columns of F hold the K dynamic factors ($N^2 \times K$), and the rows of C contain the eigenvectors ($K \times K$).

Let φ_k be the eigenvector corresponding to the k-th eigenvalue from the covariance matrix of X_{TAC}, and then:

$$\Sigma_X \varphi_k = \lambda_k \varphi_k \text{ for } k = 1, 2, \ldots, K \tag{3}$$

where Σ_X is the covariance matrix of X_{TAC} and λ_k its k-th eigenvalue.

The covariance matrix is symmetric $\Sigma_X = \Sigma_X'$ and its eigenvectors φ_k are orthogonal, that is:

$$\langle \varphi_i, \varphi_j \rangle = \varphi_i' \cdot \varphi_j = \begin{cases} 1, & i = j \\ 0, & i \neq j \end{cases}$$

Thus, it is possible to build an orthogonal matrix of size $K \times K$, defined as:

$$\Phi = \begin{bmatrix} \varphi_1 & \varphi_2 & \cdots & \varphi_K \end{bmatrix}$$

Such that:

$$\Phi'\Phi = I$$

That is:

$$\Phi^{-1} = \Phi'$$

Hence, Eq. 3 can be written as:

$$\Sigma_X \Phi = \Sigma_X \Lambda$$

where Λ is a diagonal matrix containing the eigenvalues of Σ_X.

The covariance matrix has at the most K eigenvectors associated to K nonzero eigenvalues, assuming that $K < N$, that is, the size of the image is bigger than the number of frames.

The Karhunen-Loève transform is defined as:

$$F = \left(\Phi' X_{TAC}'^* \right)' = X_{TAC}^* \Phi \tag{4}$$

where X_{TAC}^* is the group of TAC curves with the mean removed.

Post-multiplying by Φ' both sides of Eq. 4 results on:

$$X_{TAC}^* = F\Phi' \tag{5}$$

Eq. 5 is known as the inverse Karhunen-Loève transform.

When sorting the eigenvectors in descending order, according to its associated eigenvalue, the also sorted factors are obtained in agreement with the variation found in the images.

If we compare equations 2 and 5, it can be concluded that C is the transpose matrix of orthogonal eigenvectors of Σ_X, and F is a matrix containing the projections of X_{TAC}^* over an orthogonal subspace defined by C', representing regions with similar temporal behavior. Of note, in order to fully reconstruct X_{TAC}, it is necessary to add the previously removed mean.

Jimenez et al. performed the FADS to one group of healthy volunteers and three of patients with LBBB and mild and severe dilated cardiomyopathy (DCM). The first column of Fig. 19a shows the three main factors of an ERNA study performed on a subject with a normal cardiac contraction pattern. The second column plots the

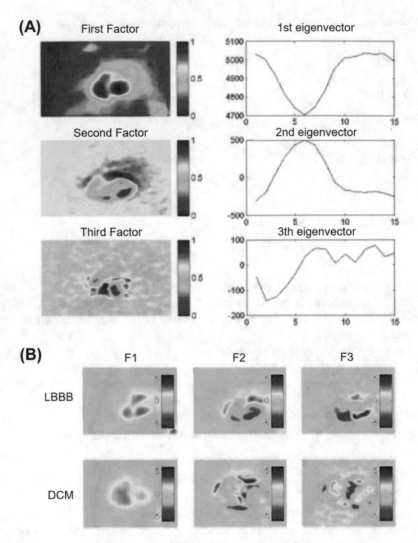

Fig. 19 Results from FADS applied to ERNA studies from a subject having a normal (**a**) contraction pattern and two (**b**) pathological cases, LBBB and DCM. (Adapted from Jiménez-Ángeles et al. [50] with permission)

eigenvectors associated to each factor. Figure 19b presents the three main factors of abnormal contraction patterns, with patients suffering from LBBB and DCM in the top and bottom rows, respectively.

Literature involving dynamic series of cardiovascular studies has reported that the first factor corresponds to the structure having the most evident motion, being this the case of the ventricular region for a healthy contraction pattern. This description holds for different abnormal patterns, as long as the ventricular mass keeps its mobility [21, 37, 38, 73].

The second factor has been associated with the atria, as it is the next structure representing an important range of motion. However, it is worth mentioning that this region is compromised in the lateral oblique view, as they hold a smaller area and thus less statistical information with regard to the motion. Results from Jimenez et al. confirm these findings for the groups of healthy and LBBB subjects. Nevertheless, this is not the case for the group of DCM patients. In spite of the fact that the second factor still shows the highest values in the auricular region, there is no strong atrioventricular contrast, as it is the case for the other two groups.

It is worth stressing that the plot of the first eigenvector follows the temporal evolution of the ventricular volume, and it rapidly increases its value up to a maxima and decreases back to the initial amplitude. On the other hand, the plot of the second eigenvector mimics the evolution of the auricular volume, starting at a maximum value that decreases rapidly and grows back to baseline.

Interpretation of the third factor has been less consistent in literature. Whereas it is associated with the behavior of the big blood vessels in a healthy subject [21], in patients, it may explain the motion variability in the atria or the ventricles, depending on the pathology [50, 51].

Rojas et al. applied FADS using ERNA-SPECT studies and compared healthy subjects with a group of patients suffering pulmonary arterial hypertension (PAH). Their results confirmed that the first two factors correspond to the behavior of the atrioventricular walls, and the third shows different behavior between the control and objective groups [75]. Figure 20 depicts a central slice on the long-axis view followed by its corresponding three main factors, for a control and pathological subjects.

In the same way as the Fourier analysis, the interpretation of results in the FADS is performed in comparison to a well-defined pattern; this is the cardiac contraction of a healthy subject. However, in contrast to the phase image analysis, FADS does not assume periodicity of the TACs but rather describes their statistical behavior. Nonetheless, an appropriate interpretation of FADS requires a full assessment of how the intensities are distributed in the main factors of a cardiac study.

From this information, a statistical model of the normal cardiac contraction pattern can be built, in order to compare it with other abnormal contraction cases. This was carried out by Jimenez et al. using ERNA studies acquired at the National Institute of Cardiology "Ignacio Chávez" (INCICh) (Mexico City), including a group of healthy controls and heart failure patients suffering from LBBB, DCM, or PAH [51]. The statistical model is described as follows:

Let $f_p = \left[F(p, 1) \; F(p, 2) \; F(p, 3) \right]'$ be a vector formed by the magnitude of the three main factors at the p-th pixel or TAC for each study. Figure 21a shows the dispersion plots of the factorial intensities from the pixels within the ventricular region for a healthy subject. Black and magenta correspond to the vectors from the left and right ventricles, respectively. Following the same notation, Fig. 21b displays the vectors corresponding to a patient with LBBB. It is evident that, meanwhile in

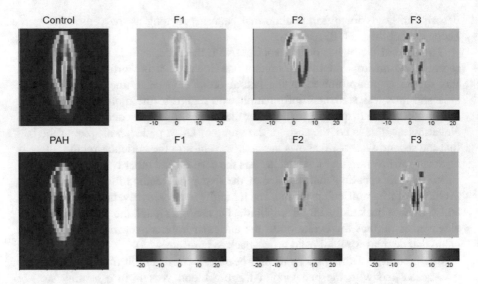

Fig. 20 Central slice from ERNA-SPECT studies in the long-axis view and their corresponding factorial images. (**a**) control participant; (**b**) PAH patient

a healthy subject the vector distribution for both ventricles in the space F1-F2-F3 overlaps, the dispersion in a LBBB patient locates different regions for the left and right ventricles. The contribution of F3 plays a fundamental role in separating both vector clusters.

Rojas et al. modeled the normal cardiac contraction pattern using planar ERNA as well as ERNA-SPECT studies. Both modalities confirmed that the vector clusters from the left and right ventricles overlap in healthy subjects; however, in PAH patients, there is a segregation between them, in this case being the second factor responsible for it [75].

Even though the FADS guarantees obtaining up to k factors from the decomposition, the three main factors concentrate more than 90% of the energy contained in the data, even for the ERNA-SPECT studies.

The same group proposed modeling the probability density function of observations $\overset{...}{f}$ from the control group as a mixture of Gaussian functions:

$$p\left(\overset{...}{f}\right) = \sum_{i=1}^{R} w_r N\left(\overset{...}{f}|\mu_r, \Sigma_r\right)$$

where R is the number of Gaussian functions and w_r is the relative weight of the r-th Gaussian in the mixture, such that:

$$\sum_{i=1}^{R} w_r = 1 \text{ and } 0 \leq w_r \leq 1$$

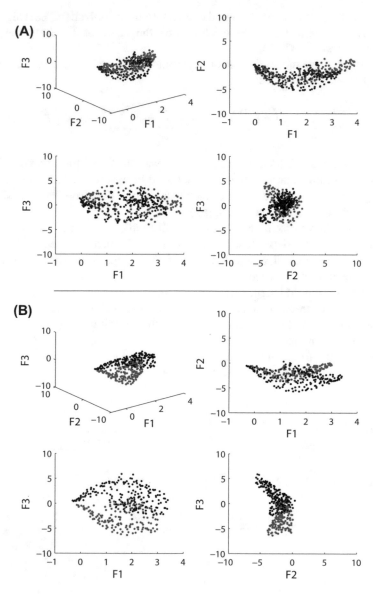

Fig. 21 Dispersion plots of the vectors corresponding to the pixels within the left (black) and right (magenta) ventricles for a control participant (**a**) and a LBBB patient (**b**). (Reproduced from Jiménez-Ángeles et al. [51] with permission)

$N\left(\overset{...}{f}|\mu_r, \Sigma_r\right)$ is the normal multivariate probability density function with parameters μ_r and Σ_r.

Parameters were estimated using the expectation-maximization (EM) algorithm. A bootstrap strategy was used to determine the value of R, using the *Bayes information criterion* (BIC).

The fitting of the model to a set of observations S is evaluated using the average log-likelihood, calculated as:

$$I_N = -\frac{1}{\|S\|}\sum_{s\in S}\log\left(p\left(\overset{...}{f}\right)\right) = -\frac{1}{\|S\|}\sum_{s\in S}\log\left(\sum_{i=1}^{R} w_r N\left(\overset{...}{f}|\mu_r, \Sigma_r\right)\right)$$

I_N was denominated "normality index," and it was computed for a group of ERNA studies from healthy subjects that were not used in the model. This value allowed defining a reference for the fitting of the model to new data from subjects showing a healthy cardiac contraction pattern [51].

Figure 22 shows box plots of the normality indices computed for all groups. T-Student tests for independent samples were performed between each patients group versus the control group. P-values of the comparisons showing statistical significance are marked in red.

Apart from the mild DCM, every other group showed difference when compared with the control group, that is, the statistical information corresponding to the dispersion of the first three dynamic factors of the control group describe a cardiac contraction pattern distinguishable from the groups of patients. This result will be considered in the following section in order to define a classification model of the severity in the cardiac contraction dyssynchrony.

Fig. 22 Box plots of the normality indices (I_N) computed for the study population. Groups with statistically significant differences are marked in red

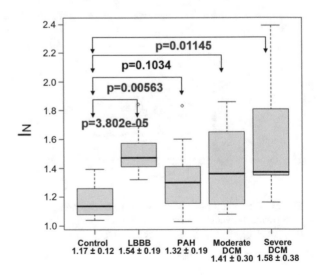

3.4 Classification of Severity in Cardiac Contraction Dyssynchrony

Jiménez-Angeles et al. reported promising findings of a model defining a healthy cardiac contraction pattern and discussed the potential use of the normality index as a criterion to qualify the severity of cardiac contraction dyssynchrony [51]. Santos-Díaz et al. proposed combining the statistical analysis from FADS with mechanical markers of cardiac contraction, obtained from the clinical file and ECG of the participants, in order to stratify the severity in cardiac contraction dyssynchrony presented by a heart failure patient and thus to create a tool to follow up patients going under CRT [77]. This section describes the proposed classification model and its main findings.

As described in Sect. 1, CRT is suggested as treatment for heart failure patients, whose inclusion criteria according to the guidelines of the NYHA are [54] as follows:

- Functional class III or IV
- Left ventricle ejection fraction (LVEF) <35%
- Duration of QRS interval >120 ms

Additionally, the type of dyssynchrony (atrio-, inter-, or intraventricular) present in the patient determines the positioning and type of resynchronizer to be implanted. Hence, it is useful to count on mechanical biomarkers to inform about the synchrony/dyssynchrony of the subject. Santos-Díaz et al. proposed the LVEF, duration of the QRS, and PR intervals as appropriate mechanical markers to describe a probable abnormality in the cardiac contraction pattern [77].

3.4.1 Study Population

Information from 75 subjects from the nuclear medicine department of the National Institute of Cardiology "Ignacio Chávez," Mexico City, was analyzed. A total of 44% of patients were healthy controls, whereas 56% were diagnosed with HF. Table 4 shows details of the studied population.

ERNA studies were acquired as described in Sect. 3.1. Three nuclear cardiologists by using the clinical information and the Fourier phase image evaluated all subjects. Experts labeled every subject as a case of absent (A), mild (M), moderate (Md), and severe (S) dyssynchrony, for both inter- and intraventricular

Table 4 Demographics of the studied population

Group	N	Age [years]
Control	33 (21 M, 12 F)	28 ± 5
Heart failure	42 (32 M, 10 F)	54 ± 15

Table 5 Number of participants described by type and degree of dyssynchrony

	Absent	Present		
		Mild	Moderate	Severe
Interventricular (LV-RV)	34	6	5	30
Intraventricular (iLV)	33	15	10	16

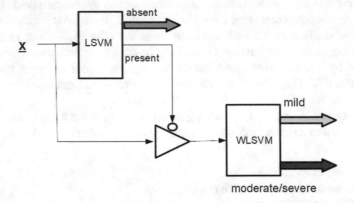

Fig. 23 Classification scheme for each type of dyssynchrony. **x**, feature vector (including LVEF, QRS, P-R, and FADS information). LSVM linear support vector machine, WLSVM weighted linear support vector machine. (Reproduced from Santos-Díaz et al. [77] with permission)

dyssynchronies. The label selected for each subject was the mode between experts. Table 5 shows the labeling process results. While there is balance between the number of healthy subjects and patients, the distribution is not homogeneous for the severity of dyssynchrony.

3.4.2 Supervised Classification

A two-layer binary classification model was proposed (Fig. 23). The first layer classified between healthy (A) and subjects presenting dyssynchrony (M, Md, or S). A positive result for dyssynchrony activates the second classification step between mild (M) and moderate-to-severe (Md-S) dyssynchronies. For both cases, linear support vector machines (LSVM) were used, implemented using R (e1071, www.r-project.org), based on the LibSVM toolbox [23, 32].

Intraventricular Dyssynchrony (iLV)

The feature vector used for this type of dyssynchrony was built using the values of LVEF, QRS, and PR and the first two eigenvectors from FADS [51].

Data vectors were separated in training and testing sets using a 70–30% proportion. Cross-validation was used to estimate the classification performance.

In the first classification layer, a LSVM for balanced samples was used, whereas for the second, they used a weighted LSVM to compensate for the imbalance between groups.

Interventricular Dyssynchrony

The feature vector used for this type of dyssynchrony was built using the values of LVEF, QRS, and PR and the second and third eigenvectors from FADS [51]. The training-testing and performance evaluation process for this layer was equivalent to that described for iLV. Table 6 shows the expected classification performance for each layer.

In a preliminary study, Santos-Díaz et al. reported the evaluation of classification models that considered the information extracted from FADS only as feature vectors and compared them with the model including information from both FADS and clinical markers. Their results were definitive, including the clinical parameters improved the classification accuracy for every configuration, for the different types of dyssynchronies evaluated [76]. Table 7 summarizes the classification results.

The proposed model achieved classification rates close to 80%. The imbalance between the number of participants with mild and moderate-to-severe dyssynchrony is noticeable; this means that most of the patients treated at the nuclear cardiology department of the INCICh are in a moderate-to-severe dyssynchrony condition. These findings invite to perform a multicenter study that includes participants across the entire dyssynchrony spectrum. Finally, having an efficient classification model in agreement with the clinical expert interpretation favors its use as a follow-up tool for patients under CRT.

Table 6 Mean accuracy +/− standard deviation for the classification models

Classes	LV-RV dyssynchrony (%)	iLV dyssynchrony (%)
Absent/present (LSVM)	96.10 ± 9.33	99.96 ± 0.82
Mild/moderate-to-severe (WLSVM)	78.17 ± 23.26	75.55 ± 23.98

LSVM linear support vector machine, *WLSVM* weighted linear support vector machine
Reproduced from Santos-Díaz et al. [77] with permission

Table 7 Classification rate for the severity in cardiac contraction dyssynchrony. Results are expressed as percentage of classification accuracy and number of hits between brackets

Dyssynchrony types	Absent	Mild	Moderate/severe	Total
iLV	100% (11/11)	50% (1/2)	90.91% (10/11)	91.67% (22/24)
LV-RV	90% (9/10)	50% (2/4)	80% (8/10)	79.17% 19/24)

4 Perspectives

Improvements in the functional evaluation of cardiac contraction mechanics include novel medical imaging methods and protocols in order to highlight information that allows increasing the sensitivity in heart failure diagnostics. With regard to cardiac imaging modalities, recent tendencies in echocardiography have led to the development of transducer arrays capable of creating tridimensional strain and velocity maps of the myocardium for diagnosis of ventricular dyssynchrony. These instruments allow measuring the strain components in all directions and many segments simultaneously, using a single acquisition. Nonetheless, tridimensional echocardiography (3DE) still suffers from lower spatial and temporal resolutions compared to its 2D variant and requires an adequate acoustic window as well. Currently, the multi-beat acquisition allows reconstructing volumetric sections of the heart with four heart beats only, but it presents artifacts in arrhythmic patients and requires breath holding. On the other hand, real-time 3DE captures motion of the entire heart in a single cardiac cycle, albeit with a lower spatiotemporal resolution. Advancement of newer technologies and data processing algorithms will solve such limitations, facilitating thus the characterization of myocardial wall deformation at higher spatial/temporal resolution and reproducibility, in order to better understand the mechanisms of ventricular dysfunction and to improve the follow-up and impact of emerging therapies [41, 49, 95].

The main challenges for gated-SPECT are dose reduction and the determination of reliable indices for ventricular dyssynchrony assessment [7]. In order to address the first limitation, the use of SPECT-cadmium zinc telluride (CZT) gamma cameras has considerably reduced the acquisition time and in consequence the radiation dose. Secondly, the ERNA has been a valuable clinical diagnostic tool leading to the establishment of robust ventricular function indices. Hence, it has recently been proposed the reprojection of 3D data into 2D images as a substitute of planar ERNA, thus taking advantage of both modalities [92].

Evaluation of the cardiac function using CMR requires a stable heart rate to facilitate ECG gating, as well as multiple breath holds in order to reduce motion artifacts; hence, acquisition times are very long. Real-time CMR attempts to reduce such inconvenient and has emerged as an alternative to the traditional methods. Recently, other more efficient acquisition protocols have been proposed. For example, through time, Generalized Autocalibrating Partial Parallel Acquisition is capable of achieving a <50 ms temporal resolution per frame. These newer methods using efficient sampling and reconstruction strategies, such as parallel imaging, compressed sensing, and retrospective navigation, allow acquiring high-resolution images without ECG gating or breath holding [1, 62].

Another development avenue has been followed in the implementation of image analysis and machine learning algorithms to assess the efficacy of pharmacological and resynchronization therapies [78].

Exploring the statistical modeling of the cardiac contraction from high-quality structural images, combining information of the contraction pattern extracted from

ERNA and ultrasound studies, and incorporating information from the clinical records are a required tendency [73]. Creation of multicenter data bases, repositories of image processing, and data analysis algorithms is another strategy that has also contributed to a better understanding of the corresponding pathologies and whose discussion leads to an expansion of new knowledge [9, 34, 78].

The fact that one-third of patients under cardiac resynchronization therapy do not respond to treatment remains; thus, it is necessary to transition toward robust prediction models that include multimodal information from signals, images, clinical records, and type of implanted device in order to predict more accurately the patient's outcome [24, 69, 89].

References

1. Aandal, G., et al. (2014). Evaluation of left ventricular ejection fraction using through-time radial GRAPPA. *Journal of Cardiovascular Magnetic Resonance, 16*(1), 79.
2. Ababneh, A. A., et al. (2000). Normal limits for left ventricular ejection fraction and volumes estimated with gated myocardial perfusion imaging in patients with normal exercise test results: Influence of tracer, gender, and acquisition camera. *Journal of Nuclear Cardiology, 7*(6), 661–668.
3. Abraham, W. T., & Hayes, D. L. (2003). Cardiac resynchronization therapy for heart failure. *Circulation, 108*(21), 2596–2603.
4. Abraham, W. T., et al. (2002). Cardiac resynchronization in chronic heart failure. *New England Journal of Medicine, 346*(24), 1845–1853.
5. Abraham, T. P., Dimaano, V. L., & Liang, H.-Y. (2007). Role of Tissue Doppler and strain echocardiography in current clinical practice. *Circulation, 116*(22), 2597–2609.
6. Aletras, A. H., et al. (1999). DENSE: Displacement encoding with stimulated echoes in cardiac functional MRI. *Journal of Magnetic Resonance, 137*(1), 247–252.
7. AlJaroudi, W., et al. (2011). Nonechocardiographic imaging in evaluation for cardiac resynchronization therapy. *Circulation. Cardiovascular Imaging, 4*(3), 334–343.
8. Auricchio, A., et al. (2004). Characterization of left ventricular activation in patients with heart failure and left bundle-branch block. *Circulation, 109*(9), 1133–1139.
9. Bai, W., et al. (2015). A bi-ventricular cardiac atlas built from 1000+ high resolution MR images of healthy subjects and an analysis of shape and motion. *Medical Image Analysis, 26*(1), 133–145.
10. Baldasseroni, S., et al. (2002). Left bundle-branch block is associated with increased 1-year sudden and total mortality rate in 5517 outpatients with congestive heart failure: A report from the Italian Network on Congestive Heart Failure. *American Heart Journal, 143*(3), 398–405.
11. Ballester-Rodés, M., et al. (2005). Base-to-apex ventricular activation: Fourier studies in 29 normal individuals. *European Journal of Nuclear Medicine and Molecular Imaging, 32*(12), 1481–1483.
12. Bank, A. J., & Kelly, A. S. (2006). Tissue Doppler imaging and left ventricular dyssynchrony in heart failure. *Journal of Cardiac Failure, 12*(2), 154–162.
13. Bauer, R., et al. (1983). In vivo/in vitro labeling of red blood cells with 99mTc. *European Journal of Nuclear Medicine, 8*(5), 218–225.
14. Belge, B., et al. (2006). Accurate estimation of global and regional cardiac function by retrospectively gated multidetector row computed tomography: Comparison with cine magnetic resonance imaging. *European Radiology, 16*(7), 1424–1433.

15. Bellenger, N. G., et al. (2000). Comparison of left ventricular ejection fraction and volumes in heart failure by echocardiography, radionuclide ventriculography and cardiovascular magnetic resonance. Are they interchangeable? *European Heart Journal, 21*(16), 1387–1396.

16. Botvinick, E., et al. (1982a). The phase image: Its relationship to patterns of contraction and conduction. *Circulation, 65*(3), 551–560.

17. Botvinick, E. H., et al. (1982b). An accurate means of detecting and characterizing abnormal patterns of ventricular activation by phase image analysis. *The American Journal of Cardiology, 50*(2), 289–298.

18. Buckberg, G., et al. (2008). Cardiac mechanics revisited: The relationship of cardiac architecture to ventricular function. *Circulation, 118*(24), 2571–2587.

19. Budge, L. P., et al. (2012). MR cine DENSE dyssynchrony parameters for the evaluation of heart failure. *JACC: Cardiovascular Imaging, 5*(8), 789–797.

20. Burkhoff, D., Oikawa, R. Y., & Sagawa, K. (1986). Influence of pacing site on canine left ventricular contraction. *American Journal of Physiology-Heart and Circulatory Physiology, 251*(2), H428–H435.

21. Cavaillolès, F., et al. (1995). Comparison between factor analysis of dynamic structures and Fourier analysis in detection of segmental wall motion abnormalities: A clinical evaluation. *The International Journal of Cardiac Imaging, 11*(4), 263–272.

22. Cerqueira, M. D., et al. (2002). Standardized myocardial segmentation and nomenclature for tomographic imaging of the heart. A statement for healthcare professionals from the Cardiac Imaging Committee of the Council on Clinical Cardiology of the American Heart Association. *Circulation, 105*(4), 539–542.

23. Chang, C.-C., & Lin, C.-J. (2011). LIBSVM: A library for support vector machines. *ACM Transactions on Intelligent Systems and Technology, 2*(3), 27:1–27:27.

24. Chen, Y., et al. (2015). Efficacy of equilibrium radionuclide angiography to predict acute response to cardiac resynchronization therapy in patients with heart failure. *Nuclear Medicine Communications, 36*(6), 610–618.

25. Chua, T., et al. (2000). Accuracy of the automated assessment of left ventricular function with gated perfusion SPECT in the presence of perfusion defects and left ventricular dysfunction: Correlation with equilibrium radionuclide ventriculography and echocardiography. *Journal of Nuclear Cardiology, 7*(4), 301–311.

26. Chuang, M. L., et al. (2000). Importance of imaging method over imaging modality in noninvasive determination of left ventricular volumes and ejection fraction: Assessment by two- and three-dimensional echocardiography and magnetic resonance imaging. *Journal of the American College of Cardiology, 35*(2), 477–484.

27. Chung, E. S., et al. (2008). Results of the predictors of response to CRT (PROSPECT) trial. *Circulation, 117*(20), 2608–2616.

28. Cleland, J. G. F., et al. (2005). The effect of cardiac resynchronization on morbidity and mortality in heart failure. *The New England Journal of Medicine, 352*(15), 1539–1549.

29. Davie, A. P., et al. (1996). Value of the electrocardiogram in identifying heart failure due to left ventricular systolic dysfunction. *BMJ, 312*(7025), 222.

30. Delgado, V., et al. (2008). Assessment of left ventricular dyssynchrony by speckle tracking strain imaging. *Journal of the American College of Cardiology, 51*(20), 1944–1952.

31. Dickstein, D. P., et al. (2008). Proton magnetic resonance spectroscopy in youth with severe mood dysregulation. *Psychiatry Research – Neuroimaging, 163*(1), 30–39.

32. Dimitriadou, E., et al. (2008). Misc functions of the Department of Statistics (e1071), TU Wien. *R Package, 1*, 5–24.

33. Dorbala, S., & Slomka, P. (2015). *Nuclear cardiology, an issue of cardiology clinics*. Elsevier Health Sciences, USA.

34. Duchateau, N., et al. (2018). Model-based generation of large databases of cardiac images: Synthesis of pathological cine MR sequences from real healthy cases. *IEEE Transactions on Medical Imaging, 37*(3), 755–766.

35. Epstein, F. H. (2007). MRI of left ventricular function. *Journal of Nuclear Cardiology, 14*(5), 729–744.

36. Farwell, D., et al. (2000). How many people with heart failure are appropriate for biventricular resynchronization? *European Heart Journal, 21*(15), 1246–1250.
37. Frouin, F., et al. (2003). FAMIS-A new tool for detecting left ventricular regional wall motion abnormalities: Clinical validation. *Journal of the American College of Cardiology, 41*(6), 452.
38. Frouin, F., et al. (2004). Factor analysis of the left ventricle by echocardiography (FALVE): A new tool for detecting regional wall motion abnormalities. *European Journal of Echocardiography, 5*(5), 335–346.
39. Galderisi, M., Cattaneo, F., & Mondillo, S. (2007). Doppler echocardiography and myocardial dyssynchrony: A practical update of old and new ultrasound technologies. *Cardiovascular Ultrasound, 5*(1), 28.
40. Ganong, W. F., & Ganong, W. (1995). *Review of medical physiology*. Norwalk, CT: Appleton & Lange.
41. Geyer, H., et al. (2010). Assessment of myocardial mechanics using speckle tracking echocardiography: Fundamentals and clinical applications. *Journal of the American Society of Echocardiography, 23*(4), 351–369.
42. Hawkins, N. M., et al. (2009). Selecting patients for cardiac resynchronization therapy. *Journal of the American College of Cardiology, 53*(21), 1944–1959.
43. Henneman, M. M., et al. (2006). Assessment of global and regional left ventricular function and volumes with 64-slice MSCT: A comparison with 2D echocardiography. *Journal of Nuclear Cardiology, 13*(4), 480–487.
44. Hesse, B., et al. (2008). EANM/ESC guidelines for radionuclide imaging of cardiac function. *European Journal of Nuclear Medicine and Molecular Imaging, 35*(4), 851–885.
45. Hoffmann, R., et al. (1996). Analysis of interinstitutional observer agreement in interpretation of dobutamine stress echocardiograms. *Journal of the American College of Cardiology, 27*(2), 330–336.
46. Hoffmann, R., et al. (1998). Standardized guidelines for the interpretation of dobutamine echocardiography reduce interinstitutional variance in interpretation. *The American Journal of Cardiology, 82*(12), 1520–1524.
47. Hoffmann, R., et al. (2005). Assessment of systolic left ventricular function: A multi-centre comparison of cineventriculography, cardiac magnetic resonance imaging, unenhanced and contrast-enhanced echocardiography. *European Heart Journal, 26*(6), 607–616.
48. Hosenpud, J. D., & Greenberg, B. H. (2007). *Congestive heart failure*. Philadelphia, PA: Lippincott Williams & Wilkins.
49. Jasaityte, R., & D'Hooge, J. (2010). Strain rate imaging: Fundamental principles and progress so far. *Imaging in Medicine, 2*(5), 547–563.
50. Jiménez-Ángeles, L., et al. (2009). Factorial phase analysis of ventricular contraction using equilibrium radionuclide angiography images. *Biomedical Signal Processing and Control, 4*(2), 149–161.
51. Jiménez-Ángeles, L., et al. (2013). Normality index of ventricular contraction based on a statistical model from FADS. *Computational and Mathematical Methods in Medicine, 2013*, 1–12.
52. Klocke, F. J., et al. (2003). ACC/AHA/ASNC guidelines for the clinical use of cardiac radionuclide imaging – executive summary. *Circulation, 108*(11), 1404–1418.
53. Knebel, F., et al. (2004). Tissue Doppler echocardiography and biventricular pacing in heart failure: Patient selection, procedural guidance, follow-up, quantification of success. *Cardiovascular Ultrasound, 2*(1), 17.
54. Levin, R., et al. (1994). *The Criteria Committee of the New York Heart Association. Nomenclature and criteria for diagnosis of diseases of the heart and great vessels* (9th ed., pp. 253–256). Boston, MA: Little, Brown & Co.
55. Lim, T. K., & Senior, R. (2006). Noninvasive modalities for the assessment of left ventricular function: All are equal but some are more equal than others. *Journal of Nuclear Cardiology, 13*(4), 445–449.

56. Lomsky, M., et al. (2008). Normal limits for left ventricular ejection fraction and volumes determined by gated single photon emission computed tomography – A comparison between two quantification methods. *Clinical Physiology and Functional Imaging, 28*(3), 169–173.
57. Malm, S., et al. (2004). Accurate and reproducible measurement of left ventricular volume and ejection fraction by contrast echocardiography: A comparison with magnetic resonance imaging. *Journal of the American College of Cardiology, 44*(5), 1030–1035.
58. McKee, P. A., et al. (1971). The natural history of congestive heart failure: The Framingham Study. *New England Journal of Medicine, 285*(26), 1441–1446.
59. Mor-Avi, V., et al. (2008). Real-time 3-dimensional echocardiographic quantification of left ventricular volumes. *JACC: Cardiovascular Imaging, 1*(4), 413–423.
60. Mukherjee, A., et al. (2015). Quantitative assessment of cardiac mechanical dyssynchrony and prediction of response to cardiac resynchronization therapy in patients with non-ischaemic dilated cardiomyopathy using equilibrium radionuclide angiography. *Europace, 18*(6), 851–857.
61. Nattel, S., et al. (2007). Arrhythmogenic ion-channel remodeling in the heart: Heart failure, myocardial infarction, and atrial fibrillation. *Physiological Reviews, 87*(2), 425–456.
62. Nayak, K. S., et al. (2015). Cardiovascular magnetic resonance phase contrast imaging. *Journal of Cardiovascular Magnetic Resonance, 17*(1), 71.
63. Nemes, A., et al. (2007). Usefulness of ultrasound contrast agent to improve image quality during real-time three-dimensional stress echocardiography. *American Journal of Cardiology, 99*(2), 275–278.
64. Nichols, K., et al. (2000). Echocardiographic validation of gated SPECT ventricular function measurements. *Journal of Nuclear Medicine, 41*(8), 1308–1314.
65. Nygren, B.-M., et al. (2014). The aortic, mitral and tricuspid annuli and their velocities: A comparative echocardiographic study. *Journal of Clinical and Experimental Cardiology, 05*(08), 1–9.
66. Onishi, T., et al. (2013). Feature tracking measurement of dyssynchrony from cardiovascular magnetic resonance cine acquisitions: Comparison with echocardiographic speckle tracking. *Journal of Cardiovascular Magnetic Resonance, 15*(1), 95.
67. Park, R. C., Little, W. C., & O'Rourke, R. A. (1985). Effect of alteration of left ventricular activation sequence on the left ventricular end-systolic pressure-volume relation in closed-chest dogs. *Circulation Research, 57*(5), 706–717.
68. Petrova, H. (2016). Some aspects related to the development, implementation and assessment of educational computer presentations. *Chemistry, 25*(4), 627–633.
69. Petrovic, M., et al. (2011). Prediction of a good response to cardiac resynchronization therapy in patients with severe dilated cardyomyopathy: Could conventional echocardiography be the answer after all? *Echocardiography, 29*(3), 267–275.
70. Pfisterer, M. E., Battler, A., & Zaret, B. L. (1985). Range of normal values for left and right ventricular ejection fraction at rest and during exercise assessed by radionuclide angiocardiography. *European Heart Journal, 6*(8), 647–655.
71. Prinzen, F. W., et al. (1990). Redistribution of myocardial fiber strain and blood flow by asynchronous activation. *American Journal of Physiology-Heart and Circulatory Physiology, 259*(2), H300–H308.
72. Puente-Barragán, A., & Jiménez-Ángeles, L. (2009). Estudio de la función ventricular con ventriculografía radioisotópica en equilibrio (VRIE). In D. Bialostozky (Ed.), *Imagenología no-invasiva cardiovascular clínica* (pp. 175–187). Permanyer.
73. Redheuil, A. B., et al. (2007). Interobserver variability in assessing segmental function can be reduced by combining visual analysis of CMR cine sequences with corresponding parametric images of myocardial contraction. *Journal of Cardiovascular Magnetic Resonance, 9*(6), 863–872.
74. Rojas Ordus, D. (2010). *Análisis de sincronía de contracción cardiaca con imágenes VRIE-SPECT*. Universidad Autónoma Metropolitana – Iztapalapa.

75. Rojas-Ordus, D., et al. (2010). Factor analysis of ventricular contraction using SPECT-ERNA images. In *32nd Annual International Conference of the IEEE EMBS* (pp. 5732–5735). IEEE.
76. Santos-Díaz, A. (2011). *Evaluación de la anormalidad en la contractilidad ventricular y la respuesta a la terapia de resincronización cardiaca.* Universidad Autónoma Metropolitana – Iztapalapa.
77. Santos-Díaz, A., et al. (2017). Automated classification of severity in cardiac Dyssynchrony merging clinical data and mechanical descriptors. *Computational and Mathematical Methods in Medicine, 2017,* 1–9.
78. Sassone, B., et al. (2018). Role of cardiovascular imaging in cardiac resynchronization therapy. *Journal of Cardiovascular Medicine, 19*(5), 211–222.
79. Saxon, L. A., & Ellenbogen, K. A. (2003). Resynchronization therapy for the treatment of heart failure. *Circulation, 108*(9), 1044–1048.
80. Schwartz, R. G., Jain, D., & Storozynsky, E. (2013). Traditional and novel methods to assess and prevent chemotherapy-related cardiac dysfunction noninvasively. *Journal of Nuclear Cardiology, 20*(3), 443–464.
81. Senni, M., et al. (1998). Congestive heart failure in the community: A study of all incident cases in Olmsted County, Minnesota, in 1991. *Circulation, 98*(21), 2282–2289.
82. Shamim, W., et al. (1999). Intraventricular conduction delay: A prognostic marker in chronic heart failure. *International Journal of Cardiology, 70*(2), 171–178.
83. Stevenson, W. G., et al. (2012). Indications for cardiac resynchronization therapy: 2011 update from the Heart Failure Society of America guideline committee. *Journal of Cardiac Failure, 18*(2), 94–106.
84. Strickberger, S. A., et al. (2005). Patient selection for cardiac resynchronization therapy: From the Council on Clinical Cardiology Subcommittee on Electrocardiography and Arrhythmias and the Quality of Care and Outcomes Research Interdisciplinary Working Group, in collaboration with the H. *Circulation, 111*(16), 2146–2150.
85. Suffoletto, M. S., et al. (2006). Novel speckle-tracking radial strain from routine black-and-white echocardiographic images to quantify dyssynchrony and predict response to cardiac resynchronization therapy. *Circulation, 113*(7), 960–968.
86. Theodoridis, S., & Koutroumbas, K. (2006). *Pattern recognition* (2nd ed.). Academic Press, USA.
87. Thompson, R. C., & Cullom, S. J. (2006). Issues regarding radiation dosage of cardiac nuclear and radiography procedures. *Journal of Nuclear Cardiology, 13*(1), 19–23.
88. Tortoledo, F. A., et al. (1983). Quantification of left ventricular volumes by two-dimensional echocardiography: A simplified and accurate approach. *Circulation, 67*(3), 579–584.
89. Tournoux, F., et al. (2015). Value of mechanical dyssynchrony as assessed by radionuclide ventriculography to predict the cardiac resynchronization therapy response. *European Heart Journal – Cardiovascular Imaging, 17*(11), 1250–1258.
90. Tseng, W.-Y. I., Su, M.-Y. M., & Tseng, Y.-H. E. (2016). Introduction to cardiovascular magnetic resonance: Technical principles and clinical applications. *Acta Cardiologica Sinica, 32,* 129–144.
91. Vallejo, E., et al. (2000). Assessment of left ventricular ejection fraction with quantitative gated SPECT: Accuracy and correlation with first-pass radionuclide angiography. *Journal of Nuclear Cardiology, 7*(5), 461–470.
92. Wackers, F. J. T. (2016). Equilibrium gated radionuclide angiocardiography: Its invention, rise, and decline and . . . comeback? *Journal of Nuclear Cardiology, 23*(3), 362–365.
93. Wilcken, D. E. L. (2015). Physiology of the normal heart. *Surgery (Oxford), 33*(2), 43–46.
94. Williams, K. A. (2005). A historical perspective on measurement of ventricular function with scintigraphic techniques: Part II – Ventricular function with gated techniques for blood pool and perfusion imaging. *Journal of Nuclear Cardiology, 12*(2), 208–215.
95. Wu, V., et al. (2014). Evaluation of diastolic function by three-dimensional volume tracking of the mitral annulus with cardiovascular magnetic resonance: Comparison with tissue Doppler imaging. *Journal of Cardiovascular Magnetic Resonance, 16*(1), 71.

96. Xiao, H. B., et al. (1996). Natural history of abnormal conduction and its relation to prognosis in patients with dilated cardiomyopathy. *International Journal of Cardiology, 53*(2), 163–170.
97. Yu, C.-M., & Hayes, D. L. (2013). Cardiac resynchronization therapy: State of the art 2013. *European Heart Journal, 34*(19), 1396–1403.
98. Yu, C.-M., et al. (2005). Predictors of response to cardiac resynchronization therapy (PROSPECT)–study design. *American Heart Journal, 149*(4), 600–605.

Pattern Recognition to Automate Chronic Patients Follow-Up and to Assist Outpatient Diagnostics

Franco Simini, Matías Galnares, Gabriela Silvera, Pablo Álvarez-Rocha, Richard Low, and Gabriela Ormaechea

Abstract Pattern Recognition detects templates in images and signals of all sorts, including what comes from sensors, communications, and information management devices. The Internet of Things is increasingly available in the population, coincidentally with soaring proportions of help-dependent people. Engineering development, in addition to consumer products and industrial goods, is offering a growing number and variety of home care and personal devices. Considering personal assistants will be in great shortage in the very near future, automatic detection of problems – both environmental and medical – is a must. The shortage of qualified helpers at home – eventually and partially substituted by automatic systems – calls for psychological adaptation and cultural changes. Once restricted to small groups of well-cared-for citizens, the benefits of medicine in the twenty-first century can be shared by ever-larger numbers of individuals, thanks to technology dissemination and Pattern Recognition. In addition to magic, placebo, and personal interaction with a physician, people of the present century will be able to rely upon technology to live longer lives. A system is first described – SIMIC – to help perform cardiac failure patient follow-up for elderly and chronic condition persons. Secondly, a personal learned assistant for physicians – PRAXIS – is described, giving him or her the support of all previous cases treated, organized in a knowledge base to reduce diagnostic errors and to increase diagnostics as well as patient management efficiency.

Keywords Pattern Recognition · ICTs · SIMIC · PRAXIS · Medical informatics

F. Simini (✉) · M. Galnares · G. Silvera · P. Álvarez-Rocha · G. Ormaechea
Universidad de la República, Montevideo, Uruguay
e-mail: simini@fing.edu.uy

R. Low
Infor-Med Medical Information Systems Inc., Woodland Hills, CA, USA

© Springer Nature Switzerland AG 2020   175
M. R. Ortiz-Posadas (ed.), *Pattern Recognition Techniques Applied to Biomedical Problems*, STEAM-H: Science, Technology, Engineering, Agriculture, Mathematics & Health, https://doi.org/10.1007/978-3-030-38021-2_8

1 Introduction

New diagnostic tools and new medical equipment allow to detect diseases at an ever earlier stage, both transmissible and non-transmissible. This trend responds to the social consensus to better health for all [2]. As a consequence of better diagnostics as well as of better treatment methods and drugs, the number of persons living with chronic conditions increases, as a consequence of their longer lives. The care and follow-up of persons with chronic conditions [1, 8, 9] demand a qualified dedication that can hardly be guaranteed to all the population in such need. In socially successful countries, national health expenditure is greater than 10% of GNP with a tendency to approach 20% in some extremely rich societies. When it comes to chronic condition care, health systems – either based on personal payment insurance or state-run intergeneration solidarity – must inevitably select certain strata of the population for full chronic care implementation. The selection criteria are often based on payment capacity of the person or of the family, reducing equity which is a goal sought by most societies. A promising alternative available today is offered by technology and Information and Communication Technologies (ICTs). We can face the design of detection, communication, and feedback devices for a standard follow-up at reasonable costs to large populations. The detection of the onset of a suspicious trend, the selection of people at home to be visited or checked upon, is the result of Pattern Recognition applications applied to clinical databases.

The millennia-old patient-physician relationship is the basis of outpatient consultation. During the visit, the patient refers a problem, and by successive iterations of a mental protocol (as well as empathy), the physician reaches a diagnostics and suggests a treatment plan. This basic medical time is captured in notes with the double goal of producing documentation and helping to reach a diagnostics with the fewer possible errors and omissions. Diagnostics is a Pattern Recognition act which must be assisted by medical informatics to increase efficiency and exactitude, fostering the natural mental capacity of the physician. The role of the physician is not substitutable but can be greatly enhanced, as we shall see in the following sections.

The first part of this chapter describes a system which interacts with a chronic condition patient collecting information to be displayed during the following visit and detects unusual patient situations, critical enough to eventually alert the health system. SIMIC is the name of this system, designed [3] for the follow-up of cardiac failure (CF) patients. Systems to locate inpatients are available [4], but apart from research on systems for monitoring thoracic fluid levels based on impedance in case an implantable pacemaker is available [13], the use of ICTs to link patient everyday life with chronic condition patients and follow-up has not been reported.

The second part refers to an assistant to the physician during outpatient consultations, putting Pattern Recognition in practice at several levels of the clinical data structure. PRAXIS, such is the name of the system, is the first product [7] to include the dynamics and essence of medical reasoning into a computer application to be used as an active assistant for subsequent patients.

2 Information and Communication Technologies in Daily Life

People now live at all times with the help of Information and Communication Technology (ICT) which becomes pervasive in all aspects of modern life. ICT helps us to remember, we talk to each other using ICTs, and we read novels and newspapers with ICT and write our memoirs, friendly or business memos, and professional texts, all with ICT. We do our banking with ICT and possibly also file our tax declarations using ICT. And yet, we still expect medical care to be given by a person, a knowledgeable and wise person, part of a healthcare system.

It is not really relevant to find out whether ICT is changing Society or whether our Society defines and produces ICT according to people's needs. If we accept the idea that ICT is actually shaped by our Society and by our goals of living conditions, our values, and standards, then there will be no surprise at the fact we use them at all times and everywhere. But, on the other hand, if ICT is seen as an invasion of alien technology which changes the way Society lives and evolves, then people will tend to keep away and reject ICT. Both points of view are present today: we tend to welcome ICT for instant communication across the globe or within the family, but we are only happy if a bone and flesh medical personnel assures us that our health is out of danger. We do not think that ICT is reliable enough to solve health problems by itself, and we call for knowledgeable personnel for quality medicine. But knowledgeable health personnel is scarce and expensive. Consequently, ICT availability and economy act as complementary pair of forces: *technology push and market pull*. ICT is entering all aspects of our lives, in particular in the health sector. Scarcity of qualified personnel and consequent high healthcare costs is giving ample margin for the development and adoption of ICT specialized tools such as Pattern Recognition-driven applications to deliver efficient and high-quality medicine, as addressed in the present chapter.

3 Personal Lifetime Analyzed by Pattern Recognition

Self-care is of paramount importance for persons with chronic conditions. Their quality of life, the progression of the disease, and ultimately their survival rate depend upon their lifestyle, nutrition, medication, and exercise, all enforced on a daily basis. Recommendations by health personnel are periodically fine-tuned and adapted, which is the essence of follow-up. Traditionally, the health system is expected to assume responsibility for follow-up, which in turn is based on patient behavioral compliance and medication adherence. The growing number of persons with chronic conditions in modern societies, as a result of the epidemiological transition, challenges the effectiveness of follow-up institutional practice. This triggers the development of automated and intelligent systems to help manage chronic conditions, such as SIMIC (for the Spanish Sistema Informático de Manejo de la Insuficiencia Cardíaca) for cardiac failure (CF) [2]. Active personalized follow-up is coded as a set of recommended clinical routines, under the formalization of systems similar to expert systems.

During the medical visit, the physician "prescribes" a set of living style recommendations adapted to each chronic condition patient. The set of indications is adapted over the life of the patient. SIMIC includes such personalized recommendations and a Pattern Recognition utility to detect abnormal behaviors to allow appropriate feedback messages to the patient and eventually to alert the healthcare team of some combination of variables, matched to a serious pattern. In much the same way a physician prescribes drugs and the administration route, SIMIC allows the medical team to prescribe an App (SIMIC downloaded into the mobile phone) fine-tuned for the stage of the chronic condition the patient is in. Such prescription is issued only to patients that are able to interact with an App, either alone or with the help of a family member or home assistant. The follow-up characteristics are recorded in the SIMIC web platform and put in practice outside the physician's office by teaching the patient to install the SIMIC App in his/her mobile. From that instant on, SIMIC App will be active asking questions and capturing data from the patient, in a very sparse and respectful way, never becoming intrusive nor insisting unnecessarily. According to the type of follow-up prescribed by the physician, SIMIC will ask questions, at statistically determined times, on lifestyle, exercise, diet, and medication, as well as general mood and family activities. This information is managed within the SIMIC App Pattern Recognition routine which in turn may eventually trigger a local recommendation or comment.

The SIMIC web-based counterpart to the mobile app acts as a backup and clinical record milieu, by which only pertinent information judged as such by the physician at some later personal visit is recorded in the Electronic Clinical Record (ECR). The multidimensional time series gathered by SIMIC over time is analyzed with Pattern Recognition techniques to detect patterns the individual person is approaching until there is a confirmation of a new "state" of health.

SIMIC is designed in such a way as to behave as an intelligent system because only meaningful combinations of data and absence of data are reported to the attention of SIMIC web users, typically the physician's clinical group. The "meaningfulness" of data is the result of Pattern Recognition Analysis. Once a pattern is detected, either alarms or recommendations are sent to the patients or to the physician, in extreme risk scenarios. Patient feedback is thus received by SIMIC central application under the form of alerts, informing health personnel that the patient is behaving in a way which is frankly outside the physician's recommendations. This alert is planned to be active only for a fraction of the cardiac failure population and can be the starting point of an active human contact seeking the patient to find out more and to see whether the health system can do anything to help. By doing so, a high quality of care can be offered, as only a fraction of the persons in follow-up will require personalized calls, to try to bring back the person's behavior to the recommendations.

Just as a wearable device is read in the physician's office after a Holter recording, in the case of SIMIC, all information gathered since the previous visit is available for the SIMIC web user (i.e., the physician) in subsequent visits. The information recorded by the patient during normal life is displayed contributing to a richly informed conversation, based on recorded facts, weight variations, exercise, and diet data. The physician may decide to include all this information, along with his or her notes, in the Electronic Clinical Record (ECR), managed by SIMIC and available for interoperability, according to the CDA standard.

4 Adoption of SIMIC by Patients and Health Staff

By ensuring a continuous communication in both directions of the CF person in his or her home environment with health personnel, unwanted situations may be detected in time to actually prevent them from doing irreversible harm. SIMIC may contribute to more efficient public health management and to lower-cost quality medicine. At any time in the past, quality medicine used to be possible provided a personal physician was devoted to the patient, who was then a monarch or other privileged person: ICTs may allow to deliver the same quality medicine of our time to large cohorts of patients. SIMIC helps to deliver personalized care for large populations, using a combination of ICT and health personnel. SIMIC is the combination of a specialized ECR platform gathering sparse patient lifestyle data, events, and physiological variables (weight, mood, diet) with counseling and alert capacity. By relieving health personnel from active follow-up tasks (such as periodic telephone calls, as part of an excellent medical care), SIMIC allows personnel to devote their professional time to CF person interview during visits. By using SIMIC, if patients show good follow-up behavior, visits may be less frequent, an additional contribution of ICT to personal intelligent healthcare. The Pattern Recognition capacity of SIMIC allows to automatically select patients to receive appropriate counseling or to be referred to physician evaluation in between scheduled outpatient visits.

Including a telematic monitoring element in the patient's life helps to promote autonomy and participation in managing the symptoms of their disease. Patients can also achieve a change of behavior in relation to their health. This telematic follow-up is indicated in chronic patients who have to learn to handle medication (insulin) or who have to adopt new life, nutritional, or mobility habits (Fig. 1).

The main goal of communication with the patient [5] is to improve the patient's health and the way the patient takes care of his or her health, and this communication can be improved in terms of the rhythm of interaction, even with a canned "intelligence" such as SIMIC. Before prescribing the device, the physician must evaluate the patient's ability to learn and to perform what is indicated and to know

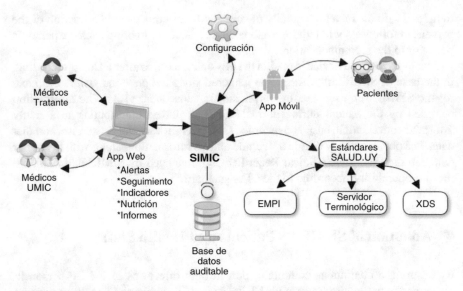

Fig. 1 General architecture of SIMIC. The App is prescribed by the physician and detects patient's lifestyle by Pattern Recognition to issue alerts and to be considered during subsequent medical visits [3]

his or her limitations. Lack of familiarity with electronic devices or rejection by the patient can hinder the decision to prescribe SIMIC.

Older patients who have resisted the use of the cell phone, or feel unable to handle ICTs, are not the best candidates for SIMIC. They will have to request the collaboration of a relative or a caregiver for its use. In these cases, it is necessary that the patient participates at all times of what is being done and thus avoid leaving him or her in a passive situation: patients must be informed of the messages, recommendations, and results of the actions that are being carried out, according to their ability to understand. Even when the patient does not use it himself, it is important that the physician addresses patient and caregiver to engage them in the task and to define the role of each one.

This contact with the patient when adopting the SIMIC follow-up technology may be the responsibility of nurses or auxiliary personnel, in the understanding that they execute a medical action. As in the past, but with different tools, it is necessary to educate the patient (including all factual information) to face life with a chronic illness.

5   Electronic Medical Records and Medical Reasoning

Taking notes, writing medical records and diagnostic decision, and following clinical guidelines are intrinsic parts of medicine. Informatics has not always been able to adapt to the essence of medical activity. Transferring data processing concepts from other fields of life such as commerce or industry, informatics has ignored some of the most specific approaches of medical mental tasks. This section presents medical reasoning as been mostly a sequence of partial Pattern Recognition phases, rejection of hypotheses, confirmation of further Pattern Recognition results, and refinements suggested by previous cases.

Informatics has taken onto itself the assistance to the medical profession and to health in general by offering simple predesigned opportunities to enter data into computers. In very much the same way as a "merchandise stock inventory" or a "pubic survey," informatics offered since the early days a way to capture medical data in sequence. Initially performed by typing clerks and helping personnel, the availability of personal computers and networks shifted data entry directly to the physician, oftentimes distorting the patient-physician relationship. No medical reasoning, no incipient Pattern Recognition, and little variations of data sequence are included in the design of most Electronic Medical Record (EMR) systems to this day.

Current EMR products are based on templates to complete the task of establishing medical records. Informatics has thus precluded the physician to create new records reusing information that he or she has previously written for other cases. Neither does the usual informatics system allow the physician to take notes in a structured way suggested by the case at hand, following a medical reasoning line of thought. The information of a patient is well-known but changes for every patient, as a result of medical knowledge, not to say medical art, applied to the case. By unifying data capture, usual informatics ignore the reality that no two physicians practice medicine in the same way [6]. Each physician has an individualized way of reaching a diagnostics and of treating patients, in addition to the fact that the cases seen follow a specific prevalence mix, very much associated with the office, the personality, or the specialization of the physician.

To start a visit, the physician always relies on a previous case, which is what happens mentally in medicine. The basic concept in medicine is one of Pattern Recognition applied to fragmented (and sometimes hidden) information. The physician can modify the initially postulated case to record accurate information about the new consultation. By modifying a similar case held in mind, the physician eventually creates a new case, which may be used for other future consultations. By giving exact meaning to such words as "case," "thought," and "concept," the following sections will describe how a Pattern Recognition approach applied to orderly data capture can record and enhance original medical reasoning.

We will see how medical informatics can help medicine by adapting to and strengthening clinical reasoning and decision-making. A Pattern Recognition approach helps to automatically suggest patient features based on previous experience of the physician.

6 PRAXIS Structure Follows Medical Reasoning

Consider PRAXIS, a medical informatics product, which makes a distinction between information recorded for each patient (clinical record) and information that reflects the physician's deliberations on the case during the visit. This distinction is not possible with traditional Electronic Medical Record (EMR) systems when filling-in simple templates, where only confirmed objective information is expected to be recorded. With such rigid templates, medicine happens in the mind of the physician, while PRAXIS tries to grab the reasoning and evolution of thought using a Pattern Recognition approach.

PRAXIS includes an EMR but goes much further to enhance the physician's reasoning. Its essence consists of easy clinical data entry and strong anticipatory help to improve medical practice. The first version of PRAXIS was developed as early as 1992 by Infor-Med Medical Information Systems Inc. in California, USA [7], as an original development stemmed loosely from a multifunctional medical system, the Perinatal Information System (SIP) initially developed for PAHO/WHO in Uruguay as early as 1985 [12].

PRAXIS is designed to create problem-oriented medical records (PROMIS) [14]. The software uses a variation of the traditional subjective-objective-assessment-prescription (SOAP) model to record insights during a medical consultation, dividing the visit information into different components.

At the start of a consultation, the physician names the clinical "case type" (CT) encountered. Later, the "case type" name can be used during searches for similar cases. The natural way for physicians to think and refer to cases is to assign nicknames such as "heavyweight housewife concerned about ecology" or "nerd male adolescent." This is modelled perfectly by a CT, very far from the template approach so common in electronic clinical record systems today, where sometimes irrelevant variables are collected disturbing medical reasoning. The Pattern Recognition put in practice by the physician is enhanced by an informatics structure that follows and records successive phases prompted by new data deduced from anamnesis, lab results, or any multiple-origin information.

The medical data structure in PRAXIS has no previous structure and creates a framework for each case as the physician puts forward a first try based on judgment and memory matching in the presence of *very* little information and successively refined into a detailed match comprising a growing data set. A point is eventually reached when it is inevitable to establish a diagnostic statement according to medical knowledge. Several otherwise captured variables are left aside, not asked for to the patient nor obtained by physical examination.

7   Representing the Knowledge Base

PRAXIS builds up the clinical knowledge base (CKB) of the physician, by accumulating his or her "case types" (CTs). The CTs reflect the way the physician reasons, describing different situations that arise when evaluating patients. This knowledge base is assembled gradually, as the physician is faced to new CTs and records them. Following the SOAP model, a CT includes not only the subjective (S) and objective (O) (of the SOAP model) from the patient but also the physician's interpretation and "decisions to treat" taken.

PRAXIS records for every CT one or more conceptual components (CCs). Figure 2 shows a generic structure representing CCs of a CT. Pharyngitis is the example, postulated initially by a physician on very weak "a priori" information. The CT is submitted to a formal SOAP approach to help the physician organize the reasoning necessary to either confirm – by successive Pattern Recognition mental matches – or reject a match with the CT altogether. The latter leads the physician to adopt another CT or to modify the one displayed on screen.

Just as a CT is composed of different conceptual components (CC), likewise, a CC consists of different conceptual elements (CEs). Figure 3 shows a structure representing different CEs of a CC. In the example included in the figure, the CC "Objective" in terms of SOAP comprises "Throat" and "Neck" CEs following the acting physician's knowledge.

Refining one step further, a CE can be further divided into different units of thought (UTs), which is the last in the hierarchy of structures. Figure 4 illustrates a structure representing the UTs of a CE. It also shows an example with UTs describing a specific throat condition.

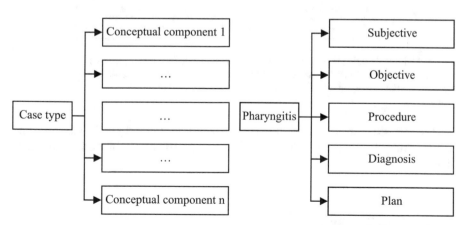

Fig. 2 Case type (CT) as defined in the mind of a physician and example of an instance guided by the SOAP medical procedure

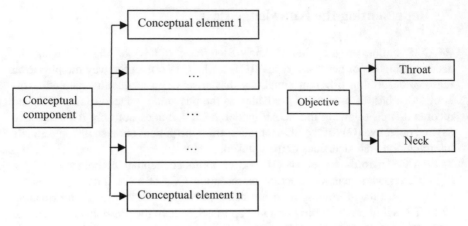

Fig. 3 Conceptual component (CC) defined as a collection of conceptual elements (CEs). The instance included here is one of the CCs described in the SOAP approach: "objective"

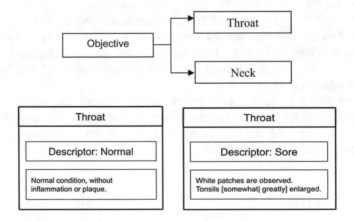

Fig. 4 Units of thought (UTs) of a conceptual element (CE). Examination of a CE "throat" (which is part of the CC "objective") can result in either a normal or sore descriptor. The UTs are "white patches" and/or "enlarged tonsils" as an alternative to "normal condition"

The tree structure of knowledge is thus based on hierarchy: UT, CE, CC, and CT. Several UTs are grouped into CEs with which to determine the CCs of a CT, as described in Fig. 5.

8 Units of Thought

The information captured by the physician consists of text strings, ordered in a tree structure. When recording a CT or "patient type," different UTs are saved, which later can be reused when recording new visits or new patients at any time. In this

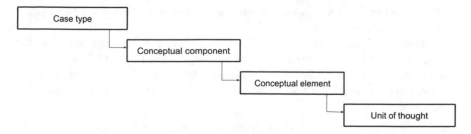

Fig. 5 Hierarchical structure of entities. The example given in previous figures is pharyngitis, objective, sore throat, tonsils enlarged, and white patches

way, previously written texts can be quickly and efficiently reused in order to capture new information, following known mental structures, away from following a rigid template. Given this ability to reuse UTs, most rewriting of text is avoided because a quick phrase matching Pattern Recognition is performed by the physician, as he or she navigates through the clinical case. Note that the same insights have most probably already been recorded for other patients at some other time. A UT is a statement that describes a basic clinical idea. A few examples are the following:

"The patient has a sore throat lasting 4 days."
"The patient denies any nasal drainage."
"She has a fever lasting 4 days."
"The patient complains of severe pain lasting 4 days."

Given that UTs model the most basic level of medical thinking, a good physician reasoning-oriented informatics system assigns some characteristics needed to give full meaning to the UT in the clinical context. An example of such characteristics would be the duration of pain. These characteristics constitute the variables used in computer science to record numeric values, times, variations according to a list of options, and gender, among others, defined by the physician user. By including such "hard" variables into the texts of UTs, a system such as PRAXIS is able to capture data to fill numeric databases, available for later statistical processing, in addition to diagnose and treat the specific case being cared for.

The same examples given above, but with the time characteristics described explicitly (in square brackets) with different durations, are the following:

"The patient has a sore throat lasting [4 days]."
"The patient has a sore throat lasting [6 days]."
"The patient has a sore throat lasting [1 week]."

If a second characteristic is defined as mild or severe pain, then the UT is "The patient complains of [mild|severe] pain lasting 4 days."

The same UT can be applied to either genders and expressed according to one of two options that does not change its meaning:

"Masculine" => "feminine"
"Man" => "woman"
"Boy" => "girl"

For example: "The boy is experiencing normal development." => "The girl is experiencing normal development."

Datum, in PRAXIS terminology, allows discrete data to be incorporated within the free text of a UT. By using Datum, the idea represented by the UT also remains unchanged but is linked to the value assigned to a variable, either text, date, numeric, or of any type. Following a markup language syntax, a text Datum can be specified as follows:

"≪patient.firstName≫ is a person who is ≪patient.age≫ years old who suffers"

To represent a UT, it is necessary to create a structure capable of storing free text, as well as referring to associated data (Datum as per PRAXIS).

Characteristics are assigned to CEs to better describe its features, in a way similar to the characteristics of a UT, but at a higher level. Each CE has an attribute indicating the default display mode of its UTs. A CE can show all its UTs by default or show only the UTs labeled as *active*, referring to the fact that they are pertinent to the UT for the given CT. The physician may label some of the UTs as *inactive*, thus modulating the CE for some particular reason. At some other time, the physician may activate any units when needed. For example, a subjective data element, which describes patient symptoms, displays its UTs as inactive. Subsequently, the physician only activates symptoms that apply to the patient being evaluated, when performing the physical examination of the patient. On the other hand, other CEs such as pharmacological indications may be displayed as active, considering that they apply to the present consultation.

CEs also have an attribute that indicates whether the element refers to the chronic information of a patient. If a CE is marked as a patient chronic information, later, it will appear on all future patient consultations. Specifically, this attribute is used to identify CEs that describe anatomical regions of patients suffering a chronic condition.

9   Interoperability

Interoperability of medical records allows EMR of a health provider to access meaningful clinical information from other health providers. Interoperability insures either only formal or clinically useful semantic information, which is of major relevance in the healthcare continuum. Several units of thought (UT) can be assembled to record a medical consultation and allow physicians to establish associations (maps) with standard codes of health terminology. The association of UTs related to diagnostics with terminology standards (such as SNOMED CT) is mandatory. For UTs that deal with medication or procedures, controlled lexical standards allow to merge statistics of different origin. The traditional medical custom of using very precise wording for each condition (often in Latin) has a natural contemporary counterpart in the adoption of standards. To associate every

UT to its code, external terminology services are used to benefit from up-to-date lexical standards.

It is clear that a good medical system will build an autonomous and personal medical knowledge base for every physician, containing all the patient records (accessible as case types) he or she has seen. It is important to note that this knowledge base is completely independent from patient EMR, strictly speaking. In other words, the system accumulates the clinical knowledge of each physician independently of the patients that the physician has evaluated.

During clinical practice, the information is saved in two databases: the patient EMR and the clinical knowledge base. Therefore, the physician can subsequently use all case types (CTs) with other patients. The system updates two data sets at the same time, the EMR and the physician clinical knowledge, which is enriched every time a new CT is recorded. In the context of several physicians working within the same clinic, the system must have a way to merge and combine the individual experience acquired with all the patients seen at the clinic by all physicians. Interestingly, the nature of clinical notes in a system like PRAXIS has the double nature of data belonging to the patients and the patient's circumstances on one side and at the same time a series of assertions, decisions, and evaluations all belonging to the physician. Never before has this double nature of clinical information been so explicitly defined as in the system we are describing here.

As a natural PRAXIS feature, the system allows physicians to exchange their "clinical knowledge base." The exchange of medical knowledge may be performed by importing CTs created by other physicians, with the explicit permission of the colleague or based on a collective office policy of sharing CTs. Imagine the rare cases seen by one of the members of a team been suggested to the rest of the physicians who have never seen one in their, maybe shorter, career.

10 Medical Records and Medical Knowledge Are Separate Entities

When recording a visit, the physician documents different elements that describe the patient's situation, and before finishing a record, the physician's clinical knowledge base is updated. However, in some specific cases, it is desirable to document a single insight about a patient, without generating clinical knowledge. By doing so, the patient EMR will be completed by adding the clinical information recorded by the physician, but no case type will be updated in the knowledge base. Note that this feature must be used sparingly since the system loses the possibility to increase medical knowledge.

The patient visit documentation method consists of taking advantage of an existing CT. The consultation of a new patient can be quickly recorded by making small changes over a similar previously written CT. The physician therefore always starts with an existing CT, which can be easily located if the physician remembers the CT name. Otherwise, it will be necessary to make use of features that help find the closest CT according to different search criteria based on associated characteristics, drugs used, or treatment plans followed in the past. This is a typically associative memory task, very common in medicine.

To simplify the CT search based on the best matches of the current patient condition, the software waits for the physician to enter the first UT, which reflects any of the aspects that may arise during the medical consultation. A Pattern Recognition approach is applied based on the first UT specified for the search. Starting with a UT, the set of CTs containing it are shown according to frequency of use. The physician may apply additional search filters to reduce the list further, in order to determine the most appropriate clinical CT to use as a basis for the new patient.

Once the physician selects the CT that is the most convenient starting point, he or she can later reuse the previous CT document. This serves the function of remembering aspects that should not be forgotten with the new patient, as well as making changes that reflect the condition of the new patient being evaluated. The CT can be quickly and conveniently changed at any time the physician realizes that it is not completely correct and wishes to search for another CT that matches more closely the current patient. To complete the consultation record, the physician should save the changes made in the selected CT. When saving a new clinical document, different situations are possible:

- It is a record of a patient with exactly the same clinical CT.
- It is a record that has improved on the same clinical CT.
- It is a record with an altogether new clinical CT.

A clinical CT is detected as a match when none of the UTs involved are different from what is true for the patient being examined. Note that a UT is considered unchanged although its characteristics (variables) may be altered such as pain duration, gender, or degree of rash spreading. Saving information about the patient consultation consists of two steps: Firstly, the clinical CT is recorded or updated in the medical knowledge base. Secondly, the patient EMR is updated, saving all the information that the physician has recorded during the medical consultation.

11 Multiple Case Types

When a patient does not arrive at a medical consultation with only one type of problem, a multiple case type (MCT) is defined. The physician may conclude multiple diagnoses for a patient during the same medical appointment. Every time the physician wants to add a new CT over another one already selected, the system is capable of combining UTs for the selected CTs. The main rules used to combine different CTs for a patient are detailed below:

- *Combination of disjoint elements*: If a CE is present in only one of the CTs, then it is added as an MCT element, since it cannot create any type of conflict.
- *Combination of common nonobjective elements*: This situation occurs when the same CE is present in more than one CT assigned to a patient, and the element does not describe patient objective information. In this situation, two possibilities are determined, according to the default display attribute of the UTs of the CE.
- *Combination of objective elements*: PRAXIS implements a more complex technique for combining CEs included in the objective component of a medical consultation. These CEs describe objective information about patient anatomical regions. In this situation, conflicts may arise if UTs are automatically combined for different CTs.

After combining several CTs to record the patient consultation, the physician may still add new UTs that detail more information about the consultation. In this context, when adding a new UT, the system will ask the physician to specify which CT should be associated to it.

PRAXIS includes features designed to ease the entry of more than one CT during the same medical consultation. In particular, it provides an option to restart the search for another CT and lets the physician specify any UT to help find the next CT. For the specific situation of MCTs, the system lets the physician consolidate the UTs from one CT to another. This involves an uncommon situation, where several CTs coexist and different UTs are combined. By consolidating a CT with another one, the first CT is augmented by adding the UTs from the second CT.

12   No Diagnostics Reached and Partial Case Types

A partial case type (PCT) is created when no diagnosis is reached in the first minutes of an appointment. For every situation in which a diagnosis is not defined, the physician can create a PCT with the CEs that he or she can confirm. In general, the patient subjective elements (what the patient says) are recorded. However, other CEs can be detailed, such as an objective element from a physical examination.

PCTs are highlighted every time they are displayed in the system. This distinction allows the physician to clearly identify the partial nature of a PCT, which makes it extremely practical when the physician wants to search for a particular PCT. By using a color distinction, it is visually easy to identify which types of cases are partial and which are not. Like other CTs, all recorded information becomes available after selecting the PCT and can be reused with future patients.

An example of a PCT would be a patient whom the physician only knows as having a bad cough. At first, no other information is available that allows the physician to conclude a differential diagnosis. In this situation, it is convenient for the physician to rely on a previous record for a case type called "Patient with a bad cough." Clearly, this PCT "Patient with a bad cough" will not have any diagnosis associated with it. However, it will present a set of symptoms, physical examinations, or other previously recorded elements related to a patient with a bad cough. In this way, the physician is supported by previous entries to determine a differential diagnosis based on his or her own past records, referring to coughs.

Therefore, every time the physician cannot define a diagnosis, he or she can search for the PCT closest to the situation of the patient being evaluated or, in the absence of one, create a new PCT. After analyzing the patient and performing the relevant physical examinations, all improvements should be saved on the created or edited PCT. Later, the physician can complete the consultation as usual, once he or she has selected a specific CT (which contains a proper diagnosis). Using PCTs until a CT is confirmed, the quality of medical records and medical practice is improved.

13   Chronic Condition and Chronic Case Type

Health monitoring allows the physician to schedule the frequency that a UT appears in a CT. In other words, this item allows the physician to plan the periodic occurrence of a UT at a certain frequency. If the health monitoring item did not exist, a UT could only be determined as *present* or *not present* in a CT. It could not model a dynamic property that considers the variable occurrence of a UTs over time in a given CT. As an example of how to use the health monitoring feature, the physician could schedule a UT for every patient with the case type "Mellitus diabetes" indicating an ophthalmologist review once a year. This item could also be used to schedule the frequency of other UTs, such as a prescription or a specific procedure.

The way in which a physician treats a chronic patient has certain specific characteristics. When a patient suffers a chronic condition, the physician has no trouble finding a diagnosis. However, the physician must be concerned with asking certain questions, verifying specific anatomical regions, and performing indications that reoccur in the patient chronic condition.

The system allows the physician to create a CT and then mark it as chronic, a chronic case types (CCT) that address the existing specifications for chronic patients. After a CT is marked as chronic, it immediately acquires a dual identity. This dual identity is the main feature of a CCT: one identity applies for the first time and the other for the same patient but at later consultations. The characteristics of these CCT identities are as follows:

First Identity   When a CCT is first associated with a patient, few differences exist between it and any other case type. Once the physician chooses the CCT, he or she will be able to reuse all the CEs of the selected CT, including the elements for the description of the patient current illness, physical examinations, indicated procedures, and prescribed medicines. As a unique characteristic, every CCT is permanently associated with the patient, and the system will assume that it is relevant to reuse it in later consultations. Therefore, after a CCT is first associated with a patient, the system will suggest that the physician consider it again every time he or she sees the same patient.

Second Identity The identity of a CCT changes significantly after the CT is first used for a patient. After being identified as a chronic condition, it is extremely important to monitor its evolution over time. The second identity of a CCT facilitates the recording of relevant information about the evolution of a chronic condition. When a CCT is expressed according to its second identity, a new conceptual component, CC, called *evolution* emerges. Under this component, the physician must specify the UT that reflects the evolution of the patient chronic condition.

From the moment the CCT acquires its second identity for a given patient, the system will no longer present the CE describing the patient current illness in that CT. The system will assume that monitoring of the chronic condition will be recorded using UTs described in the *evolution component*.

The second identity of a CCT has other key characteristics. Specifically, the second identity does not inherit the CEs that the physician has defined for the CT first identity. Therefore, when the physician uses the CT second identity for the first time, he or she will have to specify the UTs to plan the monitoring of the patient chronic condition.

After building a new CT, it is possible to mark it as chronic so that it acquires a dual identity and is assigned a highlighted color, which applies to all CCTs in the system. This distinction is quite helpful when a physician wants to locate a specific chronic case type, CCT, to use with a patient for the first time.

When a CCT is associated with a certain patient, every anatomical region that has been defined within the case type becomes a chronic region for the patient, and the system will continue to consider it in later consultations of the same patient. When working with the second identity of a chronic case type, the physician must specify the UTs that describe the monitoring planned for the patient chronic condition. In addition, it may occur that after planning a specific monitoring program, the physician wants to make a change to correct or improve the plan. Therefore, the physician can always specify new UTs that improve the monitoring.

In the context of the CCT second identity, an important feature is available when specifying a new unit of thought. When a physician adds a new unit of thought, the system immediately displays the health monitoring item. In that moment, the physician must create a new item that determines the frequency of the recently specified unit of thought. In this way, the physician can plan the monitoring of a chronic patient in a highly personalized manner, which models the reality of any tracking the physician intends to carry out.

For example, the physician could plan a monitoring program that includes a certain medicine every 2 months, an annual appointment with an ophthalmologist, and a blood test every 6 months. In this case, the physician should specify 3 units of thought with their respective monitoring items that determine the frequency of each indication. Like any other CT, when recording chronic conditions, it is quite valuable to detail a large set of controls or reminders, since the system will display the same elements for every patient with the same CCT.

14 Consultation Assistants

The system provides tools to assist physicians. The simple messaging tool and the usage reminder option were designed to help patient assistance.

Simple messaging is a communication tool that allows to send patient information to different members of the health team, who are involved in the patient care. Simple messaging allows specific patient information to be conveyed in order to notify others about certain aspects of a patient condition.

Simple messaging makes it possible to create scheduled messages that will be sent according to the moment specified by the sender. This lets the physician write a message and later decide if he or she wants to send it immediately or prefers to specify the moment that the message will be sent to the recipient.

The moments that can be defined as sending conditions are as follows:

- Immediately
- A specific date
- Next consultation
- Next time the clinical record of the patient is accessed

A physician who works with the help of receptionists can define scheduled messages for his or her helpers. Physician has features to assign tasks that receptionist must perform after the patient consultation is completed, such as coordinating a new medical appointment. Given that the physician is also one of the health team members, he or she can create messages and schedule to receive them as well. The physician can use the message tool to be reminded of certain actions he or she must complete for a patient in the future, such as before the patient next consultation.

Like any other CE, it is practical to reuse information that has already been recorded in the system. Consequently, to create new messages, it is convenient to use another similar message as a starting point. As CEs, messages can be associated with CTs. This means that when using a message in a CT, the system will remember to display the same message each time the physician evaluates another patient with the same CT.

Usage Reminder A usage reminder is an element that contributes to improving the quality of medical records, using guides and timely reminders that are displayed to the physician and are relevant to the case type being evaluated.

For every usage reminder, it is necessary to define the criteria in which the reminder will be displayed to the physician. These criteria are specified by times, conditions, and situations in which the reminder can guide, remind, or make recommendations for the physician to better address the clinical case being evaluated. Usage reminders are also quite useful in contexts where care goals are used. In general, care goals require the physician to detail additional information when treating certain patients with specific conditions. In this context, a usage reminder may be used in a timely manner to remind the physician not to forget to record the required data for fulfilling care goals. Note that to build a usage reminder, it is necessary to specify a trigger criteria, which can be simple or highly complex.

The accumulation of clinical cases in the system allows physicians to quickly register new cases, reducing writing and reading times. This method of keeping EMRs can also be used as a checklist, aiming at not forgetting important aspects that the physicians should verify with their patients, according to their own criteria. When a physician defines new chronic case type, an immediate benefit is that the tool reminds him or her of reoccurring aspects, which should be verified with the patient being evaluated but also apply to any other patient with the same CCT.

After a reasonable amount of time, the system will be able to anticipate a great part of the physician thinking regarding a reoccurring chronic patient. All the system support is based on the clinical knowledge it has learned from the physician. As a result, the physician can use his or her previous records as a checklist, to avoid forgetting certain questions and to perform all the tasks and indications relevant to the patient chronic condition. In this way, the system behaves as a clinical tool that helps physician practice better medicine and achieve quicker documentation of each consultation.

15 Conclusion

The usual application of Pattern Recognition to signal and image analysis can be extended to other complex situations such as the follow-up of chronic patients whose life and condition are described by many variables. The same applies to the multivariable patient/physician consultation and subsequent treatment plans. Pattern Recognition allows to abandon two paradigms, as we have shown in this chapter: to postpone the collection of chronic patient information to the next visit and to rely solely on physician memory to match the present case to existing ones to reach a diagnostics.

Chronic condition patients are sometimes lost to follow-up, with estimations of up to 30% lost after 2 years [11] with a deleterious effect on their quality of life as well as on health management efficiency, because these patients require repeated hospitalizations over the years, secondary to lack of adherence to treatment plans and to lifestyle recommendations. To address this problem, SIMIC adopts a completely new approach, allowing the physician to prescribe an App to the patient, just as a drug is ordered. By assigning SIMIC to be installed as an App in the cell phone of the patient (or of the person in charge), an active interaction is established gathering data during normal life at home. This information is constantly compared within the App to adaptive preestablished limits, using a Pattern Recognition approach. When the patient matches a patient type, limits are changed accordingly and proportionate action is taken by SIMIC. At first, suggestions to the patient are issued and then eventually a notification to the healthcare team. By doing so, SIMIC can help prevent losses to follow-up and may give opportunities to intervene before rehospitalization is mandatory or when facing an emotional emergency. SIMIC includes a web application which organizes follow-up data to be displayed during the next consultation [10], to help strengthen the patient-physician relationship and to deliver quality medical care.

Diagnostics and treatment plan decisions are taken by the physician at the outpatient clinic. The adoption of a Pattern Recognition approach to help diagnose, rather than data entry during the visit as a stock clerk would do with inventory merchandise, can enhance the capacity of the physician to match the current patient to previously attended cases. PRAXIS is an "infallible assistant" that remembers all details of previous similar cases, once the physician has been allowed to freely use his or her impressions and expertise to emit the first match, based on mental Pattern Recognition, i.e., deciding the case type at hand. The availability of the details, or *units of thought*, of past cases to be confirmed or dismissed is one of the key features of PRAXIS because it enhances the Pattern Recognition capacity of the physician. The still common paradigm of medical data recording in predefined sequences or templates, ordered in some professional way but seldom compatible with the reasoning stream of the physician devoted to "solve" the patient, is highly inefficient and does not help, often triggering physician rejection. PRAXIS is demonstrating with a large cohort of enthusiastic physician users that Pattern Recognition can lead to better and quicker medicine with fewer errors.

Acknowledgments The authors acknowledge the contributions of the key development team members Diego Arévalo and Miguel Pinkas of Infor-Med Inc. who have provided crucial information for the portion of this chapter devoted to PRAXIS. Gratitude is also expressed to former Universidad de la República students Alejandro Cardone, Viterbo García, and Rodrigo González in the development of SIMIC, followed by contributions by Hernán Castillo and several other medical students and software engineering students Guillermo Alvez, Michell Mamrut, and Romina Pons. Special thanks to Professor Antonio López Arredondo (Universidad de la República) for his review from the Software Engineering and Computer Science points of view, to Dr. Berta Varela for her contributions of chronic condition psychological literature and to Dr. Bruno Simini, Ospedale di Lucca, Italy, for special insight and suggestions from a clinical perspective.

References

1. Adams, J. E., & Lindemann, E. (1974). Coping with long-term disability. In G. V. Coelho, D. A. Hamburg, & J. E. Adams (Eds.), *Coping and adaptation* (pp. 127–138). New York: Basic Books.
2. Bez, M., & Simini, F. (2018, September 19–22). *Wearable devices and medical monitoring robot software to reduce costs and increase quality of care, Accepted for publication*. 7th international conference on Advances in Computing, Communications and Informatics (ICACCI 2018). Bangalore, India.
3. Cardone, A., González, R., García, V., Ormaechea, G., Álvarez-Rocha, P., & Simini, F. (2016). *Sistema de Manejo de la Insuficiencia Cardíaca (SIMIC)* (Thesis). Universidad de la República, Montevideo, Uruguay.
4. Decia, I., Farías, A., Szerman, D., Grundel, L., Briatore, D., Piñeyrúa, M., Villar, A., & Simini, F. (2016). *CAMACUA: Low cost real time risk alert and location system for healthcare environments*. CLAIB 2016. Bucaramanga, Colombia.

5. Duffy, F. D., Gordon, G. H., Whelan, G., Cole-Kelly, K., & Frankel, R. (2004). Assessing competence in communication and interpersonal skills: The Kalamazoo II report. *Academic Medicine, 79*(6), 495–507.
6. Galnares, M. (2019). *Asistente médico que aprende con la práctica para mejorar la calidad de atención* (Thesis Director Franco Simini). Universidad de la República, Uruguay.
7. Low, R. (2015). *The theory of practice: Concept processing white paper.* Infor Med Medical Systems, Inc., USA. www.praxisemr.com/white_papers.html
8. Maes, S., Leventhal, H., & de Ridder, D. (1996). Coping with chronic diseases. In M. Zeidner & N. Endler (Eds.), *Coping with chronic diseases* (pp. 221–251). New York: Wiley.
9. Moos, R. H., & Schaefer, J. A. (1984). The crisis of physical illness. In R. H. Moos (Ed.), *Coping with physical illness* (pp. 3–25). Boston: Springer.
10. Programa Salud.UY. (2018). *Presidencia de la República.* Uruguay seen May 18, 2018. https://www.agesic.gub.uy/innovaportal/v/4422/19/agesic/quees.html
11. Silvera, G., Chamorro, C., Silveira, A., Ormaechea, G., et al. (2013). Nivel de conocimiento de la enfermedad en una cohorte de pacientes con insuficiencia cardíaca. *Archivos de Medicina Interna, 35*(3), 71–75.
12. Simini, F. (1999). Perinatal information system (SIP): A clinical database in Latin America and the Caribbean. *Lancet, 354*(9172), 75.
13. Vollmann, D., Nägele, H., Schauerte, P., Wiegand, U., Butter, C., Zanotto, G., Quesada, A., Guthmann, A., Hill, M. R., & Lamp, B. (2007). Clinical utility of Intrathoracic impedance monitoring to alert patients with an implanted device of deteriorating chronic heart failure. *European Heart Journal, 28*(15), 1835–1840.
14. Weed, L. L. (1968). Medical records that guide and teach. *The New England Journal of Medicine, 278*, 593–600.

Pattern Recognition for Supporting the Replacement of Medical Equipment at Mexican Institute of Pediatrics

Tlazohtzin R. Mora-García, Fernanda Piña-Quintero, and Martha Refugio Ortiz-Posadas

Abstract The objective of this work was to develop an evaluation tool based on multi-criteria decision analysis (MCDA), for the replacement of older medical equipment installed at the National Institute of Pediatrics from Mexico, considering technical, clinical, and economic aspects. The result of such tool was an indicator called Indicator for Medical Equipment Replacement (IMER) that provides the functionality and the availability of the equipment at the medical service where it is located, in order to attend the patient's demand, as well as economical aspects. The IMER determines the medical equipment replacement priority in terms of short, medium, or long term, meaning 3, 6, or 10 years, respectively. The IMER was applied to a sample of 29 medical equipment located at three critical care units from the institute.

Keywords Medical equipment replacement · Medical equipment assessment · Replacement priority indicator · Multi-criteria decision analysis

1 Introduction

In developing countries, as Mexico, decision-makers in the healthcare sector face a global challenge of developing evidence-based methods for making decisions about medical equipment available in hospitals. The objective of this work was to develop an evaluation tool to prioritize the replacement of older medical equipment installed at the National Institute of Pediatrics from Mexico, considering three aspects: technical, clinical, and economic. Technical aspect is related to the operating status

T. R. Mora-García · M. R. Ortiz-Posadas (✉)
Electrical Engineering Department, Universidad Autónoma Metropolitana-Iztapalapa, Mexico City, Mexico
e-mail: posa@xanum.uam.mx

F. Piña-Quintero
Service of Electro-Medicine, National Institute of Pediatrics, Mexico City, Mexico

© Springer Nature Switzerland AG 2020   197
M. R. Ortiz-Posadas (ed.), *Pattern Recognition Techniques Applied to Biomedical Problems*, STEAM-H: Science, Technology, Engineering, Agriculture, Mathematics & Health, https://doi.org/10.1007/978-3-030-38021-2_9

of the device. The functionality of the medical equipment is compared to technical specifications established by the manufacturer in their service manual. Clinical aspect is about the availability of medical equipment at the medical service where it is located, in order to attend the patient's demand. Economic aspect is related to the impact of the use of medical equipment on the decision-making for the institute's own financial costs and revenues.

The evaluation tool was based on the multi-criteria decision analysis (MCDA) which is usually an ex ante evaluation tool and is particularly used for the examination of the intervention's strategic choices [1]. The application of MCDA to healthcare should be seen as a natural extension of evidence-based medicine and associated practices, such as health technology assessment [2].

The result was an indicator called Indicator for Medical Equipment Replacement (IMER) for determining the medical equipment replacement priority. It was necessary to generate a qualitative scale for interpreting its numerical result. Such scale provides three periods when the equipment must be replaced: short, medium, and long term, meaning 3, 6, and 10 years, respectively.

The indicator was applied to a sample of 28 medical equipment located at three critical care units: intensive care, neonatal intensive care, and cardiac intensive care from the institute.

2 Technological Problem

The National Institute of Pediatrics from Mexico is a tertiary public hospital with 243 beds. It has a total of 6165 medical equipment; 3699 are located in clinical and research laboratories and 2466 in healthcare areas. About 1603 equipment are less than or equal to 10 years old, 1356 are between 11 and 20 years old, and 3206 are more than 20 years old [3]. In this sense, obsolescence is a characteristic related to the medical equipment antiquity. It implies the increasingly difficult to obtain spare parts, accessories, and consumables for its correct operation, with the consequence that the equipment will stop working, so it will be necessary to acquire a new one to continue providing healthcare services to patients. The evaluation tool developed in this work provides an auxiliary criterion for making decisions about the replacement priority of the medial equipment, the rational investment of the financial resource, and, where appropriate, the acquisition of new medical equipment.

3 Methodology

The replacement model was built considering that it was a multi-criteria problem. This is then a problem with multiple and complex factors to be considered, where the choice is subject to uncertainties, risks, and different perspectives (technical, clinical, and economic), and it was this complexity that justified the use of decision analysis methods.

3.1   Multi-criteria Decision Analysis

The emphasis of multi-criteria decision analysis (MDCA) is on the judgment of the decision-making team, in establishing objectives and criteria, estimating relative importance weights, and, to some extent, judging the contribution of each option to each performance criterion [4]. In the following, we describe the steps for a multi-criteria decision analysis and what we did in each of them:

1. Establish the decision context and define the objective. The objective of this work was to provide an auxiliary criterion for making decisions about the replacement priority of the older medical equipment installed at the National Institute of Pediatrics from Mexico, based on multi-criteria decision analysis (MCDA).
2. Identify the options to be appraised for achieving the objectives. Which equipment are the oldest? Which ones are obsolete? Which equipment should be replaced first?
3. Identify the criteria to be used to compare the options. In order to know the integral performance of the medical equipment, we consider three aspects (criteria): technical, clinical, and economic. We define a set of variables considering the features of each aspect.
4. Assign weights for each of the criterion to reflect their relative importance to the decision. These weights are relative valuations assigned to each aspect considered by work team made up of biomedical engineers from the Electro-medicine Coordination (EC) and medical personnel from the critical care units considered in this study. We assign relevance factors (numerical weights) to each variable defined.
5. Combine the weights for each option to derive an overall value:

   5.1. Calculate overall weighted scores at each level in the hierarchy. A strategy followed for the health technology assessment is developing of partial indicators that calculate different measures of operational, clinical, and/or financial performance of medical equipment. We developed three partial indicators, one for each aspect considered in this study, in order to provide information about the functionality and the availability of the medical equipment, as well as economical items. We also assign relevance factors (numerical weights) to each partial indicator.
   5.2. Calculate overall weighted scores. We defined a global indicator involving not just the partial indicator but the weights for both, variables, and the partial indicators, and the result was a productivity global indicator which provides the priority for replacement particularly medical equipment called Indicator for Medical Equipment Replacement.
   5.3. Calculate overall weights for qualitative criterion. These are relative valuations of a shift between the top and bottom of the chosen scale. A qualitative scale allows an interpretation of the numerical result of the global indicator. It is usually grouped in intervals and assigned a label for their identification.

The qualitative scale for IMER was defined with three intervals, and each one indicates the period in which the equipment should be substituted: short, medium, and long term.

6. Examine the results. We analyze the numerical results of the productivity global indicator and its correspondence qualitative value, in order to find if there was correlation between these results with the performance status of the medical equipment evaluated.

4 Mathematical Model

The mathematical model involved the process of variable selection, which yields a description of the objects under study, i.e., medical equipment, and the knowledge of their relative importance (the one such variables have assigned in this case).

We defined the variables, their domains, and their weights. Afterward, we generated three partial indicators for each aspect (technical, clinical, and economic), and with these, we generated a global one, the Indicator for Medical Equipment Replacement—IMER—(Fig. 1).

4.1 Mathematical Model

Let U be a universe of objects (medical equipment), and let us consider a given finite sample $O = \{O_1, \ldots, O_m\}$ of such (descriptions of the) objects. We shall denote by $X = \{x_1, \ldots, x_n\}$ the set of variables used to study these objects. Each of these

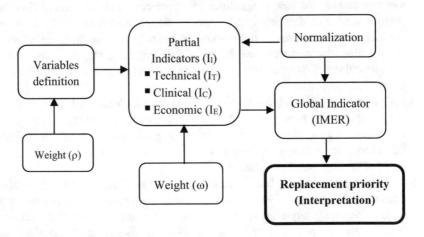

Fig. 1 Block diagram of mathematical model

variables has associated a set of admissible values (its domain of definition) M_i, $i = 1, \ldots, n$. These sets of values can be of any nature: variables can be quantitative and qualitative simultaneously. Each of these sets can contain a special symbol denoting absence of information (missing data). Thus, incomplete information about some objects is allowed. This will turn out to be a fundamental feature of this pattern recognition paradigm. By a description of an object O, we understand an n-tuple $I(O) = (x_1(O), \ldots, x_n(O))$, where $x_i: M \rightarrow M_i$, for $i = 1, \ldots, n$ are the variables used to describe it. Over M_i, no algebraic or topologic structure is assumed [5].

Definition Let the initial space representation (ISR) be the object space representation defined by the Cartesian product of M_i sets:

$$I(O) = (x_1(O), \ldots, x_n(O)) \in M_1 \times \cdots \times M_n$$

Remark There is no norm or algebraic operation over M_i defined a priori. But this does not mean that they cannot be present in ISR. Sometimes one can consider a function which does not satisfy the norm properties over M_i (over ISR) [5].

4.2 Variables

In total, 17 variables (x_i) were defined: eight for technical aspect, five for clinical aspect, and four for the economic aspect. Each variable has a qualitative domain (Q_i) used by the technical staff of the Electro-medicine Coordination (EC) for evaluating the medical equipment. However, for mathematical modelling, it was necessary to assign a quantitative domain (M_i) into [0, 1], where 1 represents the worst condition of the medical equipment and 0 the best. Each variable and each aspect have different importance, and a weighting was defined for each of them, jointly with the technical staff of the EC. The relevance for the technical aspect was $\omega_T = 0.90$, for the clinical aspect was $\omega_C = 0.50$, and for the economic aspect was $\omega_E = 0.40$. These values were designated by heuristic of the technical staff of the institute's Electro-medicine Coordination.

4.2.1 Technical Variables

Eight variables were defined for evaluating technical aspect of medical equipment (Table 1). Their qualitative (Q_i) and quantitative (M_i) domains are shown in Table 2. Observe that the variable x_5 *equipment function* defines the application and environment in which the equipment item will operate and it considers ten functions. The variable x_7 *physical risk* defines the worst-case scenario in the event of equipment malfunction. *Maintenance requirement* (x_8) describes the level and frequency of maintenance according to the manufacturer's indications or accumulated experience [6].

Table 1 Variables and weighting defined for technical aspect

		Technical aspect ($\omega_T = 0.90$)
x_i	ρ_i	Variable name
x_1	0.9	Consumables available next 5 years
x_2	0.8	Spare parts available next 5 years
x_3	0.7	Equipment age
x_4	0.6	Days of the equipment out of service
x_5	0.5	Equipment function
x_6	0.4	Equipment failure frequency
x_7	0.3	Physical risk
x_8	0.2	Maintenance requirement

Table 2 Variables (x_i) and quantitative (M_i) and qualitative domains (Q_i) for technical aspect

x_i	M_i	Q_i	x_i	M_i	Q_i
x_1	(0,1)	Yes/no	x_6	0.0	[0, 1]
x_2	(0,1)	Yes/no		0.4	[2, 4]
x_3	0.0	[1, 5]		0.6	[5, 7]
	0.4	[6, 10]		0.8	[8, 10]
	0.6	[11, 15]		1.0	>10 (failures/period evaluated)
	0.8	[16, 20]			
	1.0	[20, ∞]			
x_4	0.0	[0, 1]	x_7	0.0	No significant risk (NSR)
	0.2	[2, 3]		0.4	Patient discomfort (PD)
	0.4	[4, 5]		0.6	Inappropriate therapy or misdiagnosis (ITM)
	0.6	[6, 7]		0.8	Patient or operator injury (POI)
	0.8	[8, 10]		1.0	Patient or operator death (POD)
	1.0	>10			
x_5	0.0	Nonpatient (NP)	x_8	0.0	Minimum (M)
	0.2	Patient related and others (PR)		0.4	Less than the average (LA)
	0.3	Computer related (CR)		0.6	Average (A)
	0.4	Laboratory accessories (LR)		0.8	More than the average (MA)
	0.5	Analytical laboratory (AL)		1.0	Importants (I)
	0.6	Additional monitoring and diagnostic (AMD)			
	0.7	Surgical and intensive care monitoring (SICM)			
	0.8	Physical therapy and treatment (PTT)			
	0.9	Surgical and intensive care (SIC)			
	1.0	Life support (LS)			

Table 3 Variables defined
for clinical aspect

	Clinical aspect ($\omega_C = 0.50$)
x_i	Variable
x_9	Admissions number
x_{10}	Average stay-days
x_{11}	Use hours a day
x_{12}	Equipment number
x_{13}	Out-of-service hours

4.2.2 Clinical Variables

Five variables were defined for evaluating clinical aspect (Table 3) [7]. The variable
admissions number (x_9) refers to the number of patients that were seen in the
medical area during 2017. The variable *average stay-days* (x_{10}) indicates the average
of days that patients stay in the area. *Use hours a day* (x_{11}) quantifies the hours
that a medical equipment was used for a patient a day, considering the equipment
function. For example, in this work, we consider 31 medical equipment located at
three critical care areas of the institute. According to the World Health Organization
[8], the function of these equipment is *monitoring* for vital signs monitors (M),
physiological control for heath cradles (HC) and incubators (INC), and *support of
life* for ventilators. This equipment can be used 24 hours a day.

In the case of equipment with a *diagnostic* function such as ultrasound equipment
(US) and electrocardiograph (ECG), both are used between 30 and 90 minutes
per study per day, respectively. The estimation of the use (hours) of the medical
equipment was made jointly with the medical personnel of each area where
the equipment is used. The variable *equipment number* (x_{12}) counts the medical
equipment of the same type located at the corresponding medical area. The variable
out-of-service hours (x_{13}) considers the hours that the equipment was out of service
due to the execution of preventive maintenance routines or any failure [7]. In
addition, the constant *available hours* (k) was considered. This is the total number
of hours that the medical equipment was available in the medical service during
2017. Since the equipment can be used for 24 hours, $k = 24 (365) = 8760$ hours.

4.2.3 Economic Variables

Four variables were defined for evaluating economic aspect of medical equipment
(Table 4) [6]. Their monetary (Q_i) and quantitative (M_i) domains are shown in
Table 5. Note that Q_i is expressed in thousands of dollars (K_{USD}). With these
variables, we generate four partial indicators (I_j), and these were weighting (γ_j)
later, jointly with technical staff of the EC.

Once variables for the three aspects were defined, we proceed to generate the
partial indicators (technical, clinical, and economic) as well as the global indicator
called Indicator for Medical Equipment Replacement (IMER), as follows.

Table 4 Variables and weighting defined for economic aspects

	Economic aspect ($\omega_E = 0.40$)
x_i	Variable
x_{14}	Purchase cost
x_{15}	Maintenance cost
x_{16}	Consumables cost
x_{17}	Useful lifetime

Table 5 Variables (x_i) and quantitative (M_i) and monetary (Q_i) domains for economic aspect

x_i	M_i	Q_i (K_{USD})	x_i	M_i	Q_i (K_{USD})	x_i	M_i	Q_i (K_{USD})	x_i	$M_i = Q$
x_{14}	0.0	{0,	x_{15}	0.0	{0	x_{16}	0.0	{0	x_{17}	7
	0.1	[0–4)		0.2	(0–0.15)		0.2	(0–0.1)	(years)	10
	0.2	[0.4–10)		0.3	[0.15–0.45)		0.3	[0.1–0.25)		17
	0.3	[10–12.5)		0.4	[0.45–4)		0.4	[0.25–1)		
	0.4	[12.5–17.5)		0.5	[4–10)		0.5	[1–3.5)		
	0.5	[17.5–25)		0.6	[10–30)		0.6	[3.5–10)		
	0.6	[25–37.5)		0.7	[30–50)		0.7	[10–30)		
	0.7	[37.5–50)		0.8	[50–125)		0.8	[30–50)		
	0.8	[50–75)		0.9	[125–150]		0.9	[50–75]		
	0.9	[75–110]		1.0	>150}		1.0	>75}		
	1.0	>110}								

4.3   Partial Indicators

4.3.1   Technical Indicator (I_T)

A mathematical function (1) was developed [6] by incorporating the eight variables and its relevance factor (Table 1):

$$I_T = \frac{\sum_{i=1}^{n} \rho_i x_i}{N_T} \tag{1}$$

where x_i is the variable, $i = \{1, \ldots, 8\}$; ρ_i is the relevance factor for x_i; and N_T is the normalization parameter for obtaining the I_T result into [0, 1]. N_T is calculated by (2):

$$N_T = \sum_{i=1}^{n} \rho_i M_{i\,máx} \tag{2}$$

where $M_{i\,máx}$ is the maximum value of M_i for each variable x_i. Then, $N_T = 4.4$.

4.3.2   Clinical Indicator (I_C)

The clinical indicator [7] identifies the relationship between the demand and the supply of the medical equipment and estimates its use based on the number of patients seen in the corresponding medical service. First, two partial indicators (I_{pj}) were defined and then the clinical indicator (I_C) (Table 6).

Table 6 Economic partial indicators and their relevance

I_i	Indicator name	Mathematical function	γ_i
I_1	Purchase cost at present	$I_1 = (x_{14}\,(1 + x_3 i_R))/7.77$	0.2
I_2	Maintenance cost	$I_2 = x_{15}$	0.8
I_3	Consumables cost	$I_3 = x_{16}$	0.4
I_4	Percent of depreciation	$I_4 = (x_3/x_{17})/3$	0.6

Partial Indicator 1 Let I_{P1} be the partial indicator called *equipment demand* defined by (3):

$$I_{P1} = \frac{(x_9)\,(x_{10})\,(x_{11})}{x_{12}} \tag{3}$$

where:

x_9: Admissions number
x_{10}: Average stay-days
x_{11}: Use hours a day
x_{12}: Equipment number

Partial Indicator 2 Let I_{P2} be the partial indicator called *medical equipment supply* defined by (4):

$$I_{P2} = k - x_{13} \tag{4}$$

where:

$k = 8760$ hours
x_{13}: Out-of-service hours

Then, we defined a global *clinical indicator* (I_C) involving the two partial indicators described above. It was a ratio between the demand (I_{P1}) and the supply (I_{P2}), it allows estimating the use of the medical equipment through Eq. (5):

$$I_C = \frac{I_{P1}}{I_{P2}} \tag{5}$$

4.3.3 Economic Indicator (I_E)

For developing of economic indicator [6], four partial indicators (I_j) with their respective weighting (γ_j) were defined (Table 6). Indicator I_1 *purchase cost at present* and indicator I_4 *percent of depreciation* were normalized. We obtain $I_1 = 7.77$ because of the use of the 2017 Mexico inflation rate ($i_R = 6.77$) [9] and substituting the maximum value (1) of the variables. Therefore, to get the result into [0, 1], it was necessary to divide by 7.77. In the case of I_4, we use the years of the variables *equipment age* (x_3) and *useful lifetime* (x_{17}), and it was divided by 3 (the result of dividing the maximum value of $x_3 = 24$ and the minimum value of $x_{17} = 8$).

Then, we defined a global economic indicator (I_E) involving all partial indicators described above by Eq. (6):

$$I_E = \frac{\sum_{h=1}^{4} \gamma_j I_h}{N_E} \qquad (6)$$

where I_j is the indicator h, $h = \{1, \ldots, 4\}$; γ_j is the relevance for each indicator; and N_E is the normalization factor. N_E is calculated by (7):

$$N_E = \sum_{h=1}^{4} \gamma_h M_{h\text{máx}} \qquad (7)$$

where $M_{h\text{máx}}$ is the maximum value of M_h for each variable I_h, $h = \{1, \ldots, 4\}$. Given the maximum value for each $I_h = 1$, then, $N_E = 2$.

4.4   Indicator for Medical Equipment Replacement

The Indicator for Medical Equipment Replacement (IMER) is a mathematical function (8) that incorporates the three partial indicators and its relevance explained before:

$$\text{IMER} = \frac{\sum_{l=1}^{3} \omega_l I_l}{N_{\text{IMER}}} = \omega_T I_T + \omega_C I_C + \omega_E I_E \qquad (8)$$

$$\text{IMER} = \frac{\sum_{l=1}^{3} \omega_l I_l}{N_{\text{IMER}}} = (0.9) I_T + (0.5) I_C + (0.4) I_E$$

where:

$$N_{\text{IMER}} = \sum_{l=1}^{3} \omega_l I_{l\text{máx}} \qquad (9)$$

$$N_{\text{IMER}} = \sum_{l=1}^{3} \omega_l I_{l\text{máx}} == (0.9)(1) + (0.5)(1) + (0.4)(1) = 1.8$$

$$\text{IMER} = \frac{\sum_{l=1}^{3} \omega_l I_l}{N_{\text{IMER}}} = \frac{(0.9) I_T + (0.5) I_C + (0.4) I_E}{1.8}$$

Table 7 Qualitative scale for the Indicator for Medical Equipment Replacement (IMER)

Interval	Interpretation	Priority
(0, 0.25)	Long-term replacement (10 years)	Low
(0.25, 0.55)	Medium-term replacement (6 years)	Medium
(0.55, 1.0)	Short-term replacement (3 years)	High

4.5 Qualitative Scale for the Indicator for Medical Equipment Replacement

A qualitative scale allows an interpretation of the numerical result of the indicator. It is usually grouped in intervals and assigned a label for their identification. The qualitative scale for IMER was defined with three intervals (Table 7), and each one indicates the period in which the equipment should be substituted: short term (3 years), medium term (6 years), and long term (10 years). These intervals were defined jointly with the technical staff of the Electro-medicine Coordination of the institute. It was considered the process for purchase of medical equipment in the institute, because it needs to include the medical equipment cost in the annual budget and the purchase process takes, on average, 2 years.

5 Application of the Indicator for Medical Equipment Replacement (IMER)

The application procedure of the partial indicators (I_T, I_C, I_E) as well as the global indicator IMER in order to get the priority replacement of the medical equipment is shown in Fig. 2.

5.1 Application of the Technical Indicator (I_T)

In order to illustrate the application of the partial indicators as well as the IMER, we chose the vital signs monitor (SVM$_1$) located at ICU.

To evaluate the technical aspect of VSM1, we substitute in (1) the quantitative values of each variable and its weighting (Table 8):

$$I_{T_{MSV1}} = \frac{0.9(1)+0.8(1)+0.7(1)+0.6(0)+0.5(0.6)+0.4(0.4) + 0.3(0.6) + 0.2(0.6)}{4.4}$$

$$= 0.72$$

This result means that SVM$_1$ has a deficient functionality (remember that zero is the best condition of the medical equipment and one the worst). This is consistent with its 24 years old, and the condition about the market no longer offers the consumables or spare parts necessary for its correct operation.

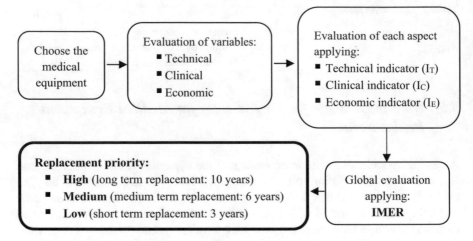

Fig. 2 Block diagram of application of the mathematical model

Table 8 Qualitative and quantitative technical description of SVM$_1$

Qualitative description								Quantitative description							
x_1	x_2	x_3	x_4	x_5	x_6	x_7	x_8	x_1	x_2	x_3	x_4	x_5	x_6	x_7	x_8
No	No	24	0	AMD	2	ITM	A	1	1	1	0	0.6	0.4	0.6	0.6

Table 9 Clinical description of VSM$_1$

x_9	x_{10}	x_{11}	x_{12}	x_{13}
468	6	24	17	2

5.2   Application of the Clinical Indicator (I_C)

The clinical description of SVM$_1$ is shown in Table 9. Note that ICU had an income of 468 patients (x_9) in 2017, with an average of 6 days-stay per patient (x_{10}). The SVM$_1$ can be used for 24 hours (x_{11}), and there are ten monitors (x_{12}). Note that VSM$_1$ was out of service for 2 hours (x_{13}).

To evaluate the clinical aspect of VSM$_1$, we substitute the quantitative values of each variable in the three indicators:

Application of I_{P1} The values of variables x_9, x_{10}, x_{11}, and x_{12} were replaced in Eq. (3):

$$I_{P1} = \frac{(468 \text{ patients}) (6 \text{ days}) (24 \text{ hours})}{(10 \text{ VSM})} \approx 3964 \text{ hours}$$

This means that each vital signs monitor was used an average of 3964 hours according to patients income in 2017.

Application of I_{P2} In the same way as for I_{P1}, the values of the constant k and x_5 were replaced in the Eq. (4):

$$I_{P2} = k - x_5 = 8760 - 2 = 8758 \text{ hours}$$

This means that VSM$_1$ was available for use for 8758 hours. This represents 99% of the total time (the crib was out of service for 2 hours during 2017).

Application of I_C To apply the I_C, the results obtained in both partial indicators were replaced in Eq. (5):

$$I_C = \frac{I_{P1}}{I_{P2}} = \frac{3964 \text{ hours}}{8758 \text{ hours}} = 0.45$$

This means that monitor SVM$_1$ has *surplus*. There is greater availability of this equipment than what is required to attend the patients who arrived at the ICU. The hours that the monitor was available for use were greater than the number of hours that the equipment was used with each patient admitted by the unit.

5.3 Application of the Economic Indicator (I_E)

To evaluate the economic aspect of the VSM$_1$, first, we applied the four partial indicators. We substituted the quantitative values of each variable and its weighting from its economic description (Table 10) in the given expression for each partial indicator (Table 6).

$$I_{1_{MSV_1}} = \frac{x_{12}(1 + x_3 i_R)}{7.77} = \frac{0.1(1 + 1(6.77))}{7.77} = 0.10$$

$$I_{2_{MSV_1}} = x_{13} = 0.20$$

$$I_{3_{MSV_1}} = x_{14} = 0.30$$

$$I_{4_{MSV_1}} = \frac{x_3 / x_{15}}{3} = \frac{24 / 10}{3} = 0.80$$

Table 10 Economic description of VSM$_1$

Monetary description					Quantitative description				
x_3	x_{14} (USD)	x_{15} (USD)	x_{16} (USD)	x_{17}	x_3	x_{14}	x_{15}	x_{16}	x_{17}
24	3777	13	123	10	0.1	1	0.2	0.3	10

In order to calculate the global economic indicator $I_{E_{MSV_1}}$, we substituted these four results in Eq. (3):

$$I_{E_{MSV_1}} = \frac{(0.2)(0.10) + (0.8)(0.20) + (0.4)(0.30) + (0.6)(0.80)}{2} = 0.39$$

This result means that the operating expenses of the monitor, from its acquisition to the present, represent almost 40%. Note that depreciation obtained the highest value ($I_{4_{MSV_1}} = 0.80$) due to its 24 years old, and considering its relevance (0.6), it has an important contribution to the result of I_E. On the other hand, although the maintenance result ($I_{2_{MSV_1}} = 0.20$) was moderate, it has the highest relevance (0.8), so its contribution into the I_E result is also important.

5.4 Application of the Indicator for Medical Equipment Replacement (IMER)

In order to do the final evaluation of the SVM$_1$, we substituted the I_T, I_C, and I_E results and its weighting (0.90 and 0.40, respectively) in the given Eq. (8) for IMER:

$$IMER_{VSM1} = \frac{\sum_{j=1}^{2} \omega_j I_j}{1.8} = \frac{\omega_T I_T + \omega_C I_C + \omega_E I_E}{1.8}$$

$$IMER_{VSM1} = \frac{(0.9)(0.72) + (0.5)(0.5) + (0.40)(0.39)}{1.8} \cong 0.59$$

According to the qualitative interpretation scale (Table 7), the result $IMER_{VSM1} = 0.59$ indicates that the monitor VSM$_1$ must be replaced in the short term (3 years), because the equipment is 24 years old; it does not have consumables or spare parts available; and its depreciation (I_4) is 80%. Therefore, it is an obsolete equipment.

6 Results

The Indicator for Medical Equipment Replacement (IMER) was applied to 29 medical equipment located at the intensive care unit (ICU), the neonatal intensive care unit (NICU), and the cardiac intensive care unit (CICU) from the institute. The medical equipment sample was distributed as follows: ten vital signs monitor (VSM), five ventilators (V), and four radiant heat cradles (RHC) at ICU; five

Table 11 Medical equipment distribution at the critical care units

Area	Equipment no.	Equipment type	Total = 28
ICU	10	Vital signs monitor (VSM)	19
	5	Ventilator (V)	
	4	Radiant heat cradle (RHC)	
NICU	5	Incubator (IC)	6
	1	Ultrasound equipment (US)	
CICU	3	Ventilator (V)	3

Table 12 Technical, clinical, economic, and IMER indicator results for VSM

Clinical area	Monitor	I_T	I_C	I_E	IMER	Replacement priority
ICU	VSM_1	0.72	0.45	0.39	0.57	3 years
	VSM_3	0.69	0.45	0.51	0.58	
	VSM_2	0.27	0.45	0.38	0.34	6 years
	VSM_4	0.29	0.45	0.43	0.37	
	VSM_5	0.23	0.45	0.38	0.32	
	VSM_6	0.29	0.45	0.43	0.37	
	VSM_7	0.65	0.45	0.38	0.53	
	VSM_8	0.65	0.45	0.37	0.32	
	VSM_9	0.27	0.45	0.38	0.34	
	VSM_{10}	0.37	0.45	0.30	0.38	

incubators, one electrocardiograph (ECG), and one ultrasonographer at NICU; and three ventilators at CICU (Table 11). It was considered the data from 2017 available in the records of the EC.

The results of the application of the three partial indicators (technical, clinical, and economic) and of the global indicator IREM in the sample of 28 medical equipment are shown below and are presented according to the type of equipment.

6.1 Vital Signs Monitors (VSM)

Ten monitors located in ICU were evaluated. The results obtained for the three partial indicators as well as for the global indicator (IMER) are shown in Table 12. Both monitors $MSV_{1,3}$ should be replaced in the short term (3 years), since they have an age of 24 and 20 years, respectively, and there are no consumables or spare parts in the market, so they are obsolete equipment. The rest of the monitors should be replaced in the medium term (6 years), since their age is between 12 and 13 years and the cost of consumables and spare parts did not exceed 50 USD during 2017.

Table 13 Technical, clinical, economic, and IMER indicator results for V

Clinical area	Ventilator	I_T	I_C	I_E	IMER	Replacement priority
ICU	V_1	0.27	0.45	0.55	0.35	6 years
	V_2	0.37	0.45	0.56	0.43	
	V_3	0.30	0.45	0.56	0.40	
	V_4	0.34	0.45	0.55	0.42	
	V_5	0.34	0.45	0.55	0.42	
UCIC	V_1	0.27	0.53	0.54	0.40	
	V_2	0.21	0.53	0.33	0.33	
	V_3	0.30	0.54	0.55	0.42	

Table 14 Technical, clinical, economic, and IMER indicator results for CCR

Clinical area	Radiant heat cradle	I_T	I_C	I_E	IMER	Replacement priority
ICU	RHC_1	0.25	0.16	0.38	0.25	6 years
	RHC_2					
	RHC_3					
	RHC_4					

6.2 Ventilators (V)

Eight ventilators were evaluated and all must be replaced in the medium term (6 years) (Table 13). In the case of the ICU, note that V_2 obtained the highest value for I_T because it is 24 years old. The others are between 12 and 13 years old, and all have consumables and spare parts available at the market. The five ventilators available in the ICU are sufficient to attend the demand of patients. Regarding the maintenance, its cost is medium because there is an external maintenance contract for these equipment. For the ventilators in the UCIC, there is an increase in the I_C value, since the demand of patients is greater and the number of available equipment is lower; therefore, there is a risk of having a deficit attention.

6.3 Radiant Heat Cradle (RHC)

Four radiant heat cradles of the ICU were evaluated. All must be replaced in the medium term (6 years) (Table 14). These equipment are between 10 and 12 years old and have consumables and spare parts available in the market. In the clinical aspect (I_C), they present a surplus, that is, they attend the demand of patients and remain available in the service. In the economic aspect (I_E), they have low costs, since the maintenance of these equipment is in charge of the technical staff of the Electro-medicine Coordination, and it was not required to buy consumables during 2017.

Table 15 Technical, clinical, economic, and IMER indicator results for INC

Clinical area	Incubator	I_T	I_C	I_E	IMER	Replacement priority
NIUC	INC$_1$	0.31	0.78	0.38	0.74	3 years
	INC$_5$					
	INC$_2$	0.25	0.78	0.38	0.68	
	INC$_3$					
	INC$_4$					

6.4 Incubators

Five incubators (INC) were evaluated. All must be replaced in 3 years (short term) (Table 15). Note that both incubators INC$_{1,5}$ obtained the highest I_T value because they are 15 years old, while the rest of them are 10 years old and have consumables and spare parts in the market. In the clinical aspect, all obtained the same value ($I_C = 0.78$) that places them in deficit. That is, the demand of patients is attended; however, the equipment are always busy. In the economic aspect (I_E), they represent a low cost, since the maintenance is carried out by the technical staff of the institute's Electro-medicine Coordination and no consumables were purchased during 2017.

6.5 Electrocardiograph (ECG)

The electrocardiograph obtained an IMER $= 0.52$, which indicates a substitution priority in the medium term (6 years). In the technical aspect, it obtained $I_T = 0.65$, due to its 15 years old and not having consumables or spare parts in the market. In the clinical aspect, it obtained $I_C = 0.32$, which means stable availability, that is, the equipment adequately attends the patient's demand and remains available in the service. Finally, in the economic aspect, it obtained $I_E = 0.32$, which indicates a low cost, since the maintenance is in charge of the technical staff of the Electro-medicine Coordination and no consumables were purchased during the evaluation period.

6.6 Ultrasound Equipment

Ultrasound equipment (US) obtained an IMER $= 0.43$ and places it with a replacement priority in the medium term (6 years). This does not agree at all with its characteristics. In the technical aspect, it obtained $I_T = 0.16$ since it is a relatively new equipment with 5 years old. In the clinical aspect, it obtained $I_C = 0.33$, which places it at a stable level, because the demand of patients is attended and the equipment remains available for use. In the economic aspect, it obtained $I_E = 0.54$, which indicates a high cost, since the maintenance is carried out by an external company and the contract cost approximately 3950.00 USD. Also during the evaluation period, the equipment required a change of the sensor, with a cost of 7950.00 USD. Although the expense of the sensor is important, it is not an accessory that has to be bought continuously.

7 Discussion

Of the total equipment evaluated, 72% must be replaced in the medium term (in a period not exceeding 6 years). These 22 medical equipment are eight vital signs monitors ($VSM_{2, 4, 5, 6, 7, 8, 9, 10}$), ten ventilators (V), and four radiant heat cradles (RHC). All these equipment are between 12 and 13 years old and still have consumables and spare parts available in the market, so they can still work for 6 more years.

The electrocardiogram (ECG) is 15 years old. It does not have spare parts and consumables available in the market and presents surplus. That is, it can attend the demand of patients and be available for a long time. For its part, the ultrasound equipment (US) is 5 years old and also has consumables and spare parts available in the market.

On the other hand, seven pieces of medical equipment (28%) must be replaced in the short term (in a period no longer than 3 years). Two vital signs monitors ($MSV_{1,3}$) are 24 years old and do not have consumables and spare parts available in the market. Five incubators (INC) are between 10 and 15 years old, and they still have consumables and spare parts available in the market. However, in this case, it should be noted that all these equipment are located in the NICU and all have deficit. There is a high demand of patients, and the equipment are at the limit of their capacity. The hospital must consider the acquisition of more medical equipment in the unit.

8 Conclusions

The IMER indicator was a very useful tool to know the status of the functionality, the clinical impact, and the expenses of the operation of the medical equipment. This information provides an auxiliary criterion for making decisions about the medical equipment that should be replaced and planning the purchase of the new one.

Later, a computational tool based on this model (IMER) will be developed with the objective to do the automatic evaluation of all the medical equipment available at the institute and promote a program of substitution of medical technology that allows to plan the acquisition for those equipment that require replacement in the short term (3 years).

The methodology presented in this work can be generalized to other contexts. That is, it can be used in the evaluation of any medical equipment, by updating the variables and their weights, according to the experience of the medical, paramedical, and technical staff of the particular clinical area. It is important to note that the model developed involves technical, clinical, and economic criteria, which makes it a complete support for decision-making, since it considers three fundamental aspects of the impact of medical equipment at medical services.

References

1. Department for Communities and Local Government. (2009). *Multi-criteria analysis: A manual.* London: Eland House Bressenden Place.
2. de Figueiredo, J., & Margarida, L. (2009). *Multicriteria model to support the replacement of hospital medical equipment* (Mestrado Integrado em Engenharia Biomédica). Instituto Superior Técnico, Portugal.
3. Piña Quintero, M. F. (2017). *Technical report from the Service of Electro-medicine.* Mexico City: National Institute of Pediatrics. (In Spanish).
4. Marsh, K., et al. (2016). Multi-criteria decision analysis to support healthcare decisions. *Value in Health, 19,* 125–137. Available at: https://www.niche1.nl/resources/content/publication_file_194_mcda_rob.pdf. Consulted 1 April 2019.
5. Ortiz-Posadas, M. R. (2017). The logical combinatorial approach applied to pattern recognition in medicine. In B. Toni (Ed.), *New trends and advanced methods in interdisciplinary mathematical science* (Series STEAM-H (Science, Technology, Engineering, Agriculture, Mathematics and Health)) (pp. 169–188). Cham/New York: Springer.
6. Mora-García, T., Piña-Quintero, F., & Ortiz-Posadas, M. (2018). Medical equipment replacement prioritization indicator using multi-criteria decision analysis. In Y. Hernández-Heredia, V. Milián-Núñez, & J. Ruiz-Shulcloper (Eds.), *Progress in artificial intelligence and pattern recognition. Lecture notes in computer science* (Vol. 11047, pp. 271–279). Cham: Springer.
7. Mora-García, T. R. (2018). *An assessment tool based on indicators for determining medical equipment replacement* (Biomedical Eng Bachelor's Thesis). Universidad Autónoma Metropolitana Iztapalapa, Mexico. (in Spanish).
8. World Health Organization. (2011). *Medical equipment maintenance program overview.* Geneva: World Health Organization. Web. 3 May 2018. http://apps.who.int/iris/bitstream/handle/10665/44587/9789241501538_eng.pdf?sequence=1
9. Central Bank of Mexico. Inflation rate. *Banxico.* Web. 4 May 2018. http://www.banxico.org.mx/portal-inflacion/index.html

Index

© Springer Nature Switzerland AG 2020
M. R. Ortiz-Posadas (ed.), *Pattern Recognition Techniques Applied to Biomedical
Problems*, STEAM-H: Science, Technology, Engineering, Agriculture,
Mathematics & Health, https://doi.org/10.1007/978-3-030-38021-2

Printed in the United States
By Bookmasters